Pseudodifferential Analysis on Conformally Compact Spaces

MEMOIRS
of the
American Mathematical Society

Number 777

Pseudodifferential Analysis on
Conformally Compact Spaces

Robert Lauter

American Mathematical Society
Providence, Rhode Island

2000 *Mathematics Subject Classification.*
Primary 58J40, 58J05, 58J35, 47G30, 46K10, 46L45.

Library of Congress Cataloging-in-Publication Data

Lauter, Robert, 1967–
  Pseudodifferential analysis on conformally compact spaces / Robert Lauter.
    p. cm. — (Memoirs of the American Mathematical Society, ISSN 0065-9266 ; no. 777)
  "Volume 163, number 777 (fourth of 5 numbers)."
  Includes bibliographical references and index.
  ISBN 0-8218-3272-7 (alk. paper)
  1. Pseudodifferential operators. 2. Compact spaces. 3. Manifolds (Mathematics) I. Title. II. Series.

QA3.A57 no. 777
[QA614.9]
510 s—dc21
[515′.7242]   2003040426

Memoirs of the American Mathematical Society

This journal is devoted entirely to research in pure and applied mathematics.

Subscription information. The 2003 subscription begins with volume 161 and consists of six mailings, each containing one or more numbers. Subscription prices for 2003 are $555 list, $444 institutional member. A late charge of 10% of the subscription price will be imposed on orders received from nonmembers after January 1 of the subscription year. Subscribers outside the United States and India must pay a postage surcharge of $31; subscribers in India must pay a postage surcharge of $43. Expedited delivery to destinations in North America $35; elsewhere $130. Each number may be ordered separately; *please specify number* when ordering an individual number. For prices and titles of recently released numbers, see the New Publications sections of the *Notices of the American Mathematical Society.*

Back number information. For back issues see the *AMS Catalog of Publications*.

Subscriptions and orders should be addressed to the American Mathematical Society, P. O. Box 845904, Boston, MA 02284-5904, USA. *All orders must be accompanied by payment.* Other correspondence should be addressed to 201 Charles Street, Providence, RI 02904-2294, USA.

Copying and reprinting. Individual readers of this publication, and nonprofit libraries acting for them, are permitted to make fair use of the material, such as to copy a chapter for use in teaching or research. Permission is granted to quote brief passages from this publication in reviews, provided the customary acknowledgment of the source is given.

Republication, systematic copying, or multiple reproduction of any material in this publication is permitted only under license from the American Mathematical Society. Requests for such permission should be addressed to the Acquisitions Department, American Mathematical Society, 201 Charles Street, Providence, Rhode Island 02904-2294, USA. Requests can also be made by e-mail to reprint-permission@ams.org.

Memoirs of the American Mathematical Society is published bimonthly (each volume consisting usually of more than one number) by the American Mathematical Society at 201 Charles Street, Providence, RI 02904-2294, USA. Periodicals postage paid at Providence, RI. Postmaster: Send address changes to Memoirs, American Mathematical Society, 201 Charles Street, Providence, RI 02904-2294, USA.

© 2003 by the American Mathematical Society. All rights reserved.
This publication is indexed in *Science Citation Index*®, *SciSearch*®, *Research Alert*®, *CompuMath Citation Index*®, *Current Contents*®/*Physical, Chemical & Earth Sciences.*
Printed in the United States of America.

∞ The paper used in this book is acid-free and falls within the guidelines established to ensure permanence and durability.
Visit the AMS home page at http://www.ams.org/

10 9 8 7 6 5 4 3 2 1   08 07 06 05 04 03

Contents

Introduction	ix
Acknowledgments:	xvi

Part 1. Fredholm theory for 0-pseudodifferential operators   1

Chapter 1. Review on basic objects of 0-geometry	3
1.1. The 0-structure algebra	3
1.2. The extended 0-blow up	4
1.3. Relation to the 0-double space X_0^2	6
1.4. The extended 0-triple space $X_{0,e}^3$	7
1.5. 0-densities	9
Chapter 2. The small 0-calculus and the 0-calculus with bounds	11
2.1. The Schwartz kernel theorem revisited	11
2.2. The small 0-calculus	11
2.3. Basic properties of the small 0-calculus	12
2.4. The 0-calculus with bounds	15
2.5. Basic properties of the 0-calculus with bounds	17
2.6. The indicial function	17
2.7. General bundles	18
Chapter 3. The b-c-calculus on an interval	19
3.1. The b-c-structure algebra	19
3.2. The b-c-double space	19
3.3. b-c-densities	21
3.4. The b-c-calculus with bounds	21
3.5. Basic properties of the b-c-calculus	22
3.6. Fredholm theory for the b-c-calculus	24
3.7. Invariance of the b-c-calculus under the $\mathbb{R}_+$-action	25
3.8. C^*-algebras of b-c-operators	26
3.9. General bundles	27
Chapter 4. The reduced normal operator	29
4.1. Definition of the reduced normal operator	29
4.2. Coordinate invariance of the reduced normal operator	30
4.3. Scale invariance of the reduced normal operator	31
4.4. Characterization of the reduced normal operator	32
4.5. Basic properties of the reduced normal operator	40
4.6. The case of 0-differential operators	42
4.7. General bundles	43

Chapter 5. Weighted 0-Sobolev spaces ... 45
  5.1. Boundedness of 0-operators of order 0 on L^2-spaces ... 45
  5.2. Weighted 0-Sobolev spaces ... 47
  5.3. General bundles ... 48

Chapter 6. Fredholm theory for 0-pseudodifferential operators ... 49
  6.1. Symbol reproducing families ... 49
  6.2. Characterization of Fredholm operators in $\Psi_0^0(X; {}^0\Omega^{1/2})$ ... 50
  6.3. Characterization of Fredholm operators in $\Psi_0^{m,k}(X; {}^0\Omega^{1/2})$ ... 53
  6.4. General bundles ... 54

Part 2. Algebras of 0-pseudodifferential operators of order 0 ... 55

Chapter 7. C^*-algebras of 0-pseudodifferential operators ... 57
  7.1. Solvable C^*-algebras ... 57
  7.2. The reduced normal operator on $S^*\partial X$ ... 57
  7.3. Extension of the symbolic structure ... 58
  7.4. The C^*-algebra generated by the reduced normal operator ... 59
  7.5. The C^*-algebra $\mathcal{B}_0^{(\mathfrak{a})}(X, {}^0\Omega^{1/2})$ ... 62
  7.6. The spectrum of the C^*-algebra $\mathcal{B}_0^{(\mathfrak{a})}(X, {}^0\Omega^{1/2})$ ... 63

Chapter 8. Ψ^*-algebras of 0-pseudodifferential operators ... 69
  8.1. Submultiplicative Ψ^*-algebras ... 69
  8.2. Ψ^*-completions of b-c- and 0-calculus ... 70

Appendix A. Spaces of conormal functions ... 73

Bibliography ... 79
  Notations ... 85
  Index ... 89

Abstract

The 0-calculus on a manifold with boundary is a micro-localization of the Lie algebra of vector fields that vanish at the boundary. It has been used by Mazzeo, Melrose to study the Laplacian of a conformally compact metric. We give a complete characterization of those 0-pseudodifferential operators that are Fredholm between appropriate weighted Sobolev spaces, and describe C^*-algebras that are generated by 0-pseudodifferential operators. An important step is understanding the so-called reduced normal operator, or, almost equivalently, the infinite dimensional irreducible representations of 0-pseudodifferential operators. Since the 0-calculus itself is not closed under holomorphic functional calculus, we construct submultiplicative Fréchet $*$-algebras that contain and share many properties with the 0-calculus, and are stable under holomorphic functional calculus (Ψ^*-algebras in the sense of Gramsch). There are relations to elliptic boundary value problems.

Received by the editor Received by the editor August 13, 2001, and in revised form March 20, 2002.

2000 *Mathematics Subject Classification.* 58J40 58J05 58J35 47G30 46K10 46L45.

Key words and phrases. 0-geometry, boundary fibration structure, solvable C^*-algebras, submultiplicative Ψ^*-algebras.

Supported in part by a scholarship of the *German Academic Exchange Service* (DAAD) within the *Hochschulsonderprogramm* III *von Bund und Ländern* and the *SFB* 478 *Geometrische Strukturen in der Mathematik* at the university of Münster.

Introduction

An open Riemannian manifold (X_0, g_0) is called a *conformally compact space* provided there exists a compact manifold X with boundary, a smooth metric h on X, and a boundary defining function $\varrho_N : X \to \overline{\mathbb{R}}_+$ of X such that (X_0, g_0) is isometric to the interior of X endowed with the metric $g := \varrho_N^{-2} h$. The boundary ∂X of X together with the conformal structure given by $h|_{\partial X}$ is known as *conformal infinity of* (X_0, g_0) [1, 21, 48, 54, 57, 70]. It is well-known that a conformally compact space (X_0, g_0) is a complete Riemannian manifold with negative sectional curvature outside a compact set. One of the basic examples of conformally compact spaces is, of course, hyperbolic space $\mathbb{H}^n := \mathbb{R}_+ \times \mathbb{R}_y^{n-1}$ with compactification given by the Poincaré-model. By taking quotients of $\mathbb{H}^n$ modulo a discrete action of a geometrically finite, torsion-free subgroup Γ of $\mathrm{PSL}(n; \mathbb{R})$ with no parabolic and elliptic elements, we obtain much more sophisticated examples of conformally compact spaces [8, 34, 61, 70], for instance, we can take Γ to be a Schottky group of hyperbolic isometries [10]. In these cases the sectional curvature equals -1 everywhere, whereas generally conformally compact spaces are not that restrictive. Conformally compact spaces, and particularly, quotients of hyperbolic space are of major interest in non-Euclidean scattering theory – see for instance [4, 5, 33, 34, 37, 47, 56, 58, 61, 75] and the references given there.

Following an approach of Mazzeo and Melrose we develop in this paper an elliptic and Fredholm theory for pseudodifferential operators modelled along the geometry of conformally compact spaces, and describe the structure of C^*- and Ψ^*-completions of the algebra of operators of order 0. The analysis on a conformally compact space (X_0, g_0) is in fact performed on its compactification X.

Let $\mathcal{V}_0(X)$ be the set of all smooth vector fields on X having bounded length with respect to the singular metric $g = \varrho_N^{-2} h$. The key observation is that $\mathcal{V}_0(X)$ is a Lie algebra that has a characterization independent of the metric g_0, namely

(0.1) $$\mathcal{V}_0(X) = \{V \in \mathcal{C}^\infty(X, TX) : V|_{\partial X} = 0\} .$$

Because of (0.1) from now on, we call following [58] the vector fields in $\mathcal{V}_0(X)$ simply *0-vector fields*, and g a 0-metric. The enveloping algebra of $\mathcal{V}_0(X)$ is said to be the algebra of *0-differential operators on* X. 0-differential operator are sometimes also called *uniformly degenerated*. Clearly, any 0-differential operator can be realized as a finite linear combination of finite products of 0-vector fields plus multiplication with a smooth function on X. In particular, the *Laplace-Beltrami* operator of a 0-metric is a second order 0-differential operator.

Let us finally note that with respect to local coordinates $(x,y) \in \overline{\mathbb{R}}_+ \times \mathbb{R}_y^{n-1}$ near the boundary of X, a 0-vector field V is locally of the form

$$V = a(x,y)x\partial_x + \sum_{j=1}^{n-1} b_j(x,y)x\partial_{y_j}$$

with coefficients a and b_j smooth up to $x = 0$. For simplicity, we say that 0-vector fields are locally generated by the vector fields $x\partial_x$ and $x\partial_y$. Consequently, a 0-differential operator A of order m is locally near the boundary of the form

(0.2) $$A = \sum_{j+|\alpha| \leq m} a_{j,\alpha}(x,y)(x\partial_x)^j(x\partial_y)^\alpha$$

with coefficients $a_{j,\alpha}$ smooth up to $x = 0$.

As shown in [53, 54, 58], building up a (pseudo)differential calculus on conformally compact spaces therefore fits into Melrose's general setting of *boundary fibration structures* [62, 66, 71]. Recall that Melrose developed a general framework for constructing pseudodifferential calculi on compact manifolds with corners starting from an appropriate Lie algebra $\mathcal{V}$ of vector fields reflecting the specific degeneracy and geometry of a given situation. The construction of the pseudodifferential calculus is in fact divided into two more or less independent parts: a construction of certain compact manifolds with corners and maps between them, and a precise analysis of conormal distributions on manifolds with corners. The latter is independent of the particular Lie algebra $\mathcal{V}$ and can be found for instance in [62, 67], whereas the first, geometric part depends heavily on the specific situation, i.e. on the choice of the Lie algebra $\mathcal{V}$. An axiomatic formulation of this approach to pseudodifferential analysis on singular configurations can be found in [66, 71]. The operators are defined by characterizing the singularities of their Schwartz kernels on a blown-up version of the cartesian product X^2. Generally, a pseudodifferential calculus in this setting consists of two parts, a *small calculus* which in fact is a filtered algebra, and families of *full calculi* that are modules over the small calculus. Many examples of boundary fibrations on manifolds with boundary have already been worked out in detail mainly by Melrose himself; let us only mention the *b-structure*, defined by the Lie algebra $\mathcal{V}_b(X)$ of all vector fields tangent to the boundary, locally given by $x\partial_x$ and ∂_y [64, 65, 68], the *cusp structure* with Lie algebra $\mathcal{V}_c(X) = \{V \in \mathcal{V}_b(X) : V\varrho_N \in \varrho_N^2 \mathcal{C}^\infty(X)\}$ with local representatives $x^2\partial_x$ and ∂_y [72], or the *edge structure* where the Lie algebra $\mathcal{V}_e(X)$ consists of all b-vector fields that are tangent to the fibers of a given fibration of the boundary [55]. Edge vector fields are locally of the form $x\partial_x$, $x\partial_y$ and ∂_z where (x,y,z) are local product coordinates near the boundary. We see that the *0-structure* is a special case of the edge structure where the fibers of the fibration are simply points.

The 0-structure corresponding to the Lie-algebra $\mathcal{V}_0(X)$ of 0-vector fields has been first considered by Mazzeo [53, 54] and Mazzeo, Melrose [58]. They use the associated pseudodifferential calculus, the *0-calculus*, to study the resolvent of the Laplacian of a conformally compact metric. Because of its relation to conformally compact spaces, the 0-calculus is an essential tool in scattering theory on asymptotically hyperbolic and conformally compact spaces, see, for instance, [4, 5, 33, 37, 56, 58, 61] and the references given there. Moreover, the 0-calculus plays an important role in the linearization of the Yamabe equation [1], the Einstein equation [21], and the monopole equation [87] on conformally compact spaces.

Finally, note that for an arbitrary differential operator P of order m on X, the operator $P_m := \varrho_N^m P$ is a 0-differential operator of order m on X. This simple fact relates the 0-calculus to the theory of (elliptic) boundary value problems. We will come back to this later on. As observed above, the 0-structure is a special case of the edge structure, and some applications in geometry in fact require not only the 0-structure but the full edge structure [31, 74].

There are various approaches to a pseudodifferential calculus containing parametrices of elliptic 0- or more generally, elliptic edge operators in the literature. Mazzeo [53, 54, 55, 57] and Mazzeo, Melrose [58] developed a full pseudodifferential calculus containing the inverses and Fredholm inverses of elliptic 0-differential operators in the spirit of characterizing lifted Schwartz kernels and boundary fibration structures. On the other hand, 0- resp. edge differential operators can also be realized as the differential operators within a general pseudodifferential calculus on differentiable groupoids [76, 79]. A Fredholm theory for those pseudodifferential operators has been developed in joint work with Nistor [46] and Monthubert, Nistor [43] using C^*-closures of operators of order 0 and $-\infty$. The main step to make the general results work for the 0-calculus is to integrate the Lie algebroid associated to the Lie algebra $\mathcal{V}_0(X)$ of 0-vector fields [78]. An explicit construction of the groupoid corresponding to the 0- resp. edge calculus can be found in [46]. From a more pedestrian point of view, Råde [86] studied Semi-Fredholm properties of 0-differential operators using a priori estimates; similarly, Lee studied Fredholm 0-differential operators on conformally compact Einstein manifolds [49]. Senichkin [95] constructed a pseudodifferential calculus on manifolds with smooth geometric edges containing the 0-differential calculus as a special case; Fredholm properties are discussed within a C^*-algebraic framework by characterizing all irreducible representations [96]. Finally, Schulze and his collaborators embedded edge differential operators (up to a weight factor) as the upper left corner of a 2×2 block matrix pseudodifferential calculus of Boutet de Monvel type, automatically including the theory of elliptic boundary value problems: we refer the interested reader to [90, 91, 92, 93] and the references given there. The pseudodifferential calculus is defined formally similar to the calculus on closed manifolds but using appropriate operator valued symbols instead. A detailed comparison of these various calculi still needs to be done and is the content of a forthcoming paper. In this paper, we restrict ourselves to the Mazzeo-Melrose approach.

However, it is a common feature of all these different approaches to a Fredholm theory for 0-differential operators that two invariants are necessary to characterize the Fredholmness resp. compactness of a given 0-differential operator, the *principal symbol* reflecting the symbolic part, and the *normal operator* taking care of the behavior at the boundary. Since the symbolic computation is almost identical to the case of closed manifolds, let us focus on the normal operator of a 0-differential operator A. It corresponds, in fact, to a family of model problems on the half-space $\overline{\mathbb{R}}_+ \times \mathbb{R}_y^{n-1}$; roughly speaking, the normal operator $\mathcal{N}(A)(q)$ at a point $q \in \partial X$ is obtained by freezing the coefficients at $q \in \partial X$: if A is of the form (0.2) and $q \in \partial X$ corresponds to $(0,0) \in \overline{\mathbb{R}}_+ \times \mathbb{R}_y^{n-1}$, then the normal operator $\mathcal{N}(A)(q)$ is given by

$$(0.3) \qquad \mathcal{N}(A)(q) = \sum_{j+|\alpha| \leq m} a_{j,\alpha}(0,0)(x\partial_x)^j (x\partial_y)^\alpha, \, (x,y) \in \overline{\mathbb{R}}_+ \times \mathbb{R}_y^{n-1},$$

i.e. by a 0-type problem on the model half space $\overline{\mathbb{R}}_+ \times \mathbb{R}_y^{n-1}$. The Fredholm property of A is equivalent to the invertibility of the principal symbol and the invertibility of the normal operators $\mathcal{N}(A)(q)$ between appropriate, weighted Sobolev spaces on $\overline{\mathbb{R}}_+ \times \mathbb{R}_y^{n-1}$ for all $q \in \partial X$. However, it is important to stress that the weight has to be uniform in $q \in \partial X$. From a more analytical point of view, Fourier transform along $y \in \mathbb{R}_y^{n-1}$ transforms the normal operator $\mathcal{N}(A)(q)$ into a family of Bessel type ordinary differential operators

$$(0.4) \qquad \mathcal{N}(A)(q)(\eta) = \sum_{j+|\alpha|\leq m} a_{j,\alpha}(0,0) i^{|\alpha|} (x\partial_x)^j x^{|\alpha|} \eta^\alpha , \, \eta \in \mathbb{R}_\eta^{n-1},$$

that is easier to deal with. The scale invariance of (0.4) even allows to restrict oneself to the case $|\eta| = 1$.

In the present paper, following a suggestion of Melrose we study a modified version of the normal operator that is equally well-adapted to pseudodifferential operators. Indeed, using the same kind of symmetry of the normal operator that allows to reduce the 0-type problem (0.3) to the Bessel type problem (0.4) we end up with a family $\mathcal{N}^\nu_{\varrho_N}(A)(\widehat{\eta})$ of b-c-operators on the interval $M = [0,1]$ that should be thought of a the radial compactification of the half-axis $\overline{\mathbb{R}}_+$. Here, a b-c-operator is an operator that is of b-type near 0 and of cusp type near 1. We call this reduced family of operators, that is now parametrized by $\widehat{\eta} \in S^*\partial X$, the *reduced normal operator*. The reduced normal operator of a 0-pseudodifferential operator is in fact the main tool to understand not only the Fredholm theory in the 0-calculus in the first part of the paper but also the structure of the C^*-algebra generated by the algebra of 0-operators of order 0 in the second part.

The main result of the first part of the paper gives a characterization of Fredholmness and compactness for 0-pseudodifferential operators. Note that a 0-pseudodifferential operator $A \in \Psi_0^m(X; {}^0\Omega^{1/2})$ acts boundedly on a scale of weighted Sobolev spaces $\varrho_N^{\mathfrak{a}} H_0^s(X, {}^0\Omega^{1/2})$, $\mathfrak{a}, s \in \mathbb{R}$.

THEOREM: *Let $\mathfrak{a} \in \mathbb{R}$ be an arbitrary weight.*
  (a) *The operator $A : \varrho_N^{\mathfrak{a}} H_0^s(X, {}^0\Omega^{1/2}) \to \varrho_N^{\mathfrak{a}} H_0^{s-m}(X, {}^0\Omega^{1/2})$ is compact if and only if ${}^0\sigma^{(m)}(A) = 0$ and $\mathcal{N}^\nu_{\varrho_N}(A)(\widehat{\eta}) = 0$ for all $\widehat{\eta} \in S^*\partial X$.*
  (b) *The operator $A : \varrho_N^{\mathfrak{a}} H_0^s(X, {}^0\Omega^{1/2}) \to \varrho_N^{\mathfrak{a}} H_0^{s-m}(X, {}^0\Omega^{1/2})$ is Fredholm if and only if ${}^0\sigma^{(m)}(A)(\zeta) \neq 0$ for all $\zeta \in {}^0S^*X$, and the reduced normal operator*

$$(0.5) \qquad \mathcal{N}^\nu_{\varrho_N}(A)(\widehat{\eta}) : \varrho_0^{\mathfrak{a}} \varrho_1^s H_{b,c}^s(M, {}^{b,c}\Omega^{1/2}) \longrightarrow \varrho_0^{\mathfrak{a}} \varrho_1^{s-m} H_{b,c}^{s-m}(M, {}^{b,c}\Omega^{1/2})$$

  is invertible for all $\widehat{\eta} \in S^*\partial X$.

This theorem extends naturally to 0-pseudodifferential operators acting between sections of arbitrary vector bundles on X. Analogous results for 0-differential operators can be found for instance in [**54, 55, 58, 86**] but using the normal operator instead. We emphasize that (0.5) is a condition for *all $\widehat{\eta} \in S^*\partial X$*. In fact, the set of possible weights $\mathfrak{a} \in \mathbb{R}$ is determined by the condition that the *indicial function* $I_A \in \mathcal{H}(\mathbb{C}, \mathcal{C}^\infty(\partial X))$ of A satisfies $I_A\left(\xi - i(\mathfrak{a} + \frac{n-1}{2})\right)(q) \neq 0$ for all $q \in \partial X$ and all $\xi \in \mathbb{R}$. The indicial function of A coincides with the indicial operator of the b-part of the reduced normal operator of A, and detects in some sense what of the b-calculus is hidden in the 0-calculus.

Let us elaborate a little bit more on the invertibility of the reduced normal operator which is in fact a serious condition. For simplicity, let us assume that we have $m = 0$ and the operator A acts on sections of the associated 0-half density bundle, a notion that is explained in the text. The reader can think on scalar operators up to an appropriate choice. Then the ellipticity of A, i.e. the invertibility of the principal 0-symbol ${}^0\sigma^{(0)}(A)$ implies that for each $q \in \partial X$ there exists a discrete set $D(q) \subseteq \mathbb{C}$ such that the reduced normal operator $\mathcal{N}^\nu_{\varrho_N}(A)(\widehat{\eta})$ is Fredholm as an operator $\varrho_0^{\mathfrak{a}} L^2(M, {}^{b,c}\Omega^{1/2}) \to \varrho_0^{\mathfrak{a}} L^2(M, {}^{b,c}\Omega^{1/2})$ for all $\mathfrak{a} \notin \text{Im}(D(q))$ and all $\widehat{\eta} \in S^*\partial X$ with $\pi(\eta) = q$. However, for A to induce a Fredholm operator $A : \varrho_N^{\mathfrak{a}} L^2(X, {}^0\Omega^{1/2}) \to \varrho_N^{\mathfrak{a}} L^2(X, {}^0\Omega^{1/2})$ we need to find a weight $\mathfrak{a} \in \mathbb{R}$ such that $\mathfrak{a} \notin \text{Im}(D(q))$ uniformly in $q \in \partial X$. Even then, the reduced normal operator is only a smooth family of Fredholm operators, so the index bundle is, for instance, an important obstruction for the invertibility of the reduced normal operator. Fortunately, there are many important examples coming from applications in geometry that show that the reduced normal operator can indeed be invertible; see, for instance [54, 58].

As noted above, there are intimate relations between the 0-calculus and boundary value problems, the invertibility of the reduced normal operator of $P_m = \varrho_N^m P$, for instance, is closely related to the so-called condition of Lopatinskij-Shapiro in boundary value problems for the differential operator P. The corresponding analysis is performed in the language of Boutet de Monvel calculus for boundary value problems [7], i.e. in the form of 2×2-block matrices containing additionally boundary conditions of trace and potential type. However, given that there exists a uniform weight $\mathfrak{a}$ such that the reduced normal operator $\mathcal{N}^\nu_{\varrho_N}(P_m)$ of an elliptic 0-differential operator P_m is a family of Fredholm operators between weighted Sobolev spaces, it can be completed by adding boundary terms to an invertible family of block-matrix operators if and only if the *Atiyah-Bott obstruction* vanishes, i.e. if the associated index bundle $\text{ind}\,\mathcal{N}^\nu_{\varrho_N}(P_m) \in K(S^*\partial X)$ actually belongs to $\pi^* K(\partial X)$ where $\pi : S^*\partial X \to \partial X$ is the canonical projection. To explore the deep and demanding details of boundary value problems and Boutet de Monvel's algebra within a pseudodifferential calculus of 0- resp. edge operators is beyond the scope of this paper. We mainly restrict to the upper left corner and refer the reader to [18, 90, 91, 93, 94] and the references given there.

In the second part of the paper we consider the algebra of 0-operators of order 0 from the point of view of topological algebras. We mainly focus on two properties of a given unital algebra $\mathcal{A}$:

(a) Is $\mathcal{A}$ a Ψ^*-*algebra*, i.e. is there a Fréchet topology on $\mathcal{A}$ and a C^*-algebra $\mathcal{B}$ such that $\mathcal{A}$ can be realized as a symmetric, continuously embedded subalgebra of $\mathcal{B}$ satisfying

$$\mathcal{A} \cap \mathcal{B}^{-1} = \mathcal{A}^{-1}.$$

(b) Is there a submultiplicative Fréchet topology on $\mathcal{A}$, i.e. is there a Fréchet topology on $\mathcal{A}$ that is generated by a sequence of *submultiplicative* seminorms, i.e. seminorms p that satisfy in addition $p(AB) \leq p(A)p(B)$ for all $A, B \in \mathcal{A}$.

The notion of Ψ^*-algebras has been introduced by Gramsch [22] in connection with a perturbation theory for algebras of singular integral and pseudodifferential operators. The Ψ^*-property has quite a few remarkable properties, for instance, for the structure of the sets of idempotent and relatively regular elements [23]. Almost by definition, Ψ^*-algebras are closed under holomorphic functional calculus, thus, the

K-theory of a Ψ^*-algebra coincides with that of its C^*-closure [6, 12]. The important point to note here is that while the rigid setting of C^*-algebras is an adequate tool to understand continuous and Fredholm properties, the more flexible Fréchet topology of a Ψ^*-algebra allows the treatment of various $\mathcal{C}^\infty$-phenomena within a functional analytic framework, for instance, elliptic regularity and propagation of singularity type results can be understood within a Ψ^*-setting [24, 28]. Many algebras occurring naturally in pseudodifferential analysis are known to be Ψ^*-algebras, see, for instance, [2, 3, 13, 30, 88, 89, 100]; however, the Ψ^*-property does not hold for the small b-calculus [40]. For more properties and examples of Ψ^*-algebras we refer to the brief review in [40].

Note that submultiplicativity of Fréchet algebras plays an important role in connection with Oka-principle and non-abelian complex cohomology [25, 27], K-theory [80], and the bivariant K-theory kk for so-called m-algebras [14]. In the context of pseudodifferential operators and boundary value problems submultiplicativity could be established for a number of interesting algebras, we refer the reader to [2, 26, 29, 30] and the references therein.

Though there is a natural Fréchet topology on the algebra $\Psi_0^0(X; {}^0\Omega^{1/2})$, we see exactly as for the b-calculus [40] that there is no topology on $\Psi_0^0(X; {}^0\Omega^{1/2})$ making it into a topological algebra with an *open* group of invertible elements, in particular, $\Psi_0^0(X; {}^0\Omega^{1/2})$ is neither closed under holomorphic functional calculus in any of the C^*-algebras $\mathcal{L}(\varrho_N^\mathfrak{a} L^2(X, {}^0\Omega^{1/2}))$ nor it has the Ψ^*-property. As for the second property, at present we do not know whether the Fréchet topology on $\Psi_0^0(X; {}^0\Omega^{1/2})$ can be generated by a system of *submultiplicative* seminorms.

We start the second part of the paper by studying the C^*-algebra $\mathcal{B}_0^{(\mathfrak{a})}(X, {}^0\Omega^{1/2})$ generated by the algebra $\Psi_0^0(X; {}^0\Omega^{1/2})$ with respect to the C^*-norm induced by the C^*-algebra $\mathcal{L}(\varrho_N^\mathfrak{a} L^2(X, {}^0\Omega^{1/2}))$ in detail. We show in particular, that $\mathcal{B}_0^{(\mathfrak{a})}(X, {}^0\Omega^{1/2})$ is solvable of length 2 in the sense of Dynin [17], hence it is nuclear and of type I. Moreover, we give a complete list of all irreducible representations of $\mathcal{B}_0^{(\mathfrak{a})}(X, {}^0\Omega^{1/2})$ up to unitary equivalence and characterize the Jacobson topology on the spectrum of $\mathcal{B}_0^{(\mathfrak{a})}(X, {}^0\Omega^{1/2})$. Recall that the spectrum together with the Jacobson topology can be thought of as an adequate non-commutative replacement of the space of maximal ideals in the commutative world. Spectra of C^*-algebras of pseudodifferential operators with various kinds of discontinuities in the symbols have been computed before by Plamenevskij and Senichkin [81, 82, 84, 85]. We expect the C^*-algebra of pseudodifferential operators on manifolds with smooth edges considered by Senichkin [95, 96] to be the closest to our algebra of 0-operators. However, since the definition of the operators is completely different, a comparison of the two C^*-algebras requires some more efforts that will be done elsewhere. In joint work with Monthubert and Nistor we proved in [43] the existence of composition series for C^*-algebras of pseudodifferential operators on certain differentiable groupoids. Together with the construction in [46], these results apply to the 0- and the edge calculus.

Finally, we describe certain submultiplicative Ψ^*-completions of the algebra $\Psi_0^0(X; {}^0\Omega^{1/2})$. This approach has been suggested to us by Gramsch in the context of the b-calculus [41]. Instead of looking at the algebra $\Psi_0^0(X; {}^0\Omega^{1/2})$ directly, we consider submultiplicative Ψ^*-algebras $\mathcal{A}_0^{(\mathfrak{a})}(X, {}^0\Omega^{1/2})$ between $\Psi_0^0(X; {}^0\Omega^{1/2})$ and $\mathcal{B}_0^{(\mathfrak{a})}(X, {}^0\Omega^{1/2})$. This answers in particular the question which $\mathcal{C}^\infty$-properties of a

0-pseudodifferential operator of order 0 are preserved under holomorphic functional calculus. The construction of the algebras $\mathcal{A}_0^{(\mathfrak{a})}(X, {}^0\Omega^{1/2})$ follows the techniques outlined in full generality in [**30, 40**], and already used in [**41**] for the b-calculus.

Roughly speaking, we use commutator methods as introduced in [**3, 16**] for the characterization of pseudodifferential operators. The main advantage of these methods is that the resulting algebras are automatically submultiplicative.

We expect the results of this paper to extend readily to the edge calculus; details will be given elsewhere.

The paper is organized as follows: In CHAPTER 2 we introduce the basic objects of the 0-boundary fibration structure. It is worth mentioning at this point that we replaced the 0-double space X_0^2 of [**54, 58**] by a slightly modified space obtained by blowing up an additional piece of X_0^2; this extended 0-double space $X_{0,e}^2$ has several technical advantages; for instance, the extended 0-front face naturally fibers over $[-1, 1] \times \partial X$ with typical fiber diffeomorphic to a half-sphere of dimension $n - 1$. Note that introducing this new double space does not change the small 0-calculus, however, we gain a little bit of flexibility in the definition of the full 0-calculus. The use of the extended 0-double space has been suggested to us by Melrose. CHAPTER 3 is devoted to the definition of the 0-calculus. In CHAPTER 4 we recall the basic properties of the b-c-calculus on M; the properties of the C^*-algebra generated by b-c-operators of order 0 can also be found in that Chapter. The reduced normal operator is the subject of CHAPTER 5. The characterization of the reduced normal operator for arbitrary 0-operators is the most important technical part of the paper and probably of interest on its own. CHAPTER 6 deals with the scale of weighted 0-Sobolev spaces that is necessary for the Fredholm theory of 0-operators of arbitrary order. The Fredholm theory for the 0-calculus can be found in CHAPTER 7. The second part of the paper starts with CHAPTER 8 where the structure of the C^*-algebras $\mathcal{B}_0^{(\mathfrak{a})}(X, {}^0\Omega^{1/2})$ is studied in detail. Finally, CHAPTER 9 is devoted to submultiplicative Ψ^*-completions of the algebra $\Psi_0^0(X; {}^0\Omega^{1/2})$. The paper contains an APPENDIX about certain properties of conormal functions on $\mathbb{R}^N$ that are necessary for the characterization of the reduced normal operator of 0-operators of order $-\infty$. These properties are perhaps well-known to experts.

To fix some notations, let Z be a manifold with corners in the sense of [**68**]. We write $\mathcal{M}_k(Z)$ for the family of *boundary faces of codimension* k, $k = 0, \ldots, \dim Z$. The elements of $\mathcal{M}_1(Z)$ are also called *boundary hyperfaces*; by definition of a manifold with corner, for each $H \in \mathcal{M}_1(Z)$, there is a smooth function $\varrho_H : Z \to \overline{\mathbb{R}}_+$ satisfying $H = \{\varrho_H = 0\}$ and $d\varrho_H \neq 0$ at H; ϱ_H is said to be a *defining function* for H.

Let $\mathcal{V}_b(Z)$ be the Lie algebra of all smooth vector fields on Z that are tangent to all boundary hyperfaces of Z. Then there exists a smooth vector bundle ${}^bTZ$ over Z together with a smooth map $j^b : {}^bTZ \to TZ$ such that $\mathcal{V}_b(Z) = j^b \mathcal{C}^\infty(Z, {}^bTZ)$ [**64, 68**]; we call ${}^bTZ$ the *b-tangent bundle of Z*.

For a smooth vector bundle E over Z, we write $\dot{\mathcal{C}}^\infty(Z, E)$ for the space of smooth sections of E that vanish to infinite order at all boundary faces, $\dot{\mathcal{C}}_c^\infty(Z, E)$ for the space of those $f \in \dot{\mathcal{C}}^\infty(Z, E)$ that are compactly supported, and, finally, $\mathcal{C}^{-\infty}(Z, E) := \dot{\mathcal{C}}_c^\infty(Z, E^* \otimes \Omega(Z))'$ for the space of *extendible distributions on Z with coefficients in E*. Here, E^* is the bundle dual to E, and $\Omega(Z)$ is the bundle of full 1-*densities* on Z. Moreover, $S^{[m]}(E)$ stands for the space of smooth functions $E \setminus \{0\} \to \mathbb{C}$ that are positively homogeneous of degree $m \in \mathbb{C}$. For Z closed, we denote by $\mathscr{S}(E)$ the space of smooth functions $E \to \mathbb{C}$ that are rapidly decreasing at infinity. It is easily seen that the Schwartz space $\mathscr{S}(E)$ corresponds via pull-back to the space $\dot{\mathcal{C}}^\infty(\overline{E})$ of smooth functions vanishing to infinite order at the boundary $\mathbb{S}(E) := (E \setminus \{0\})/\mathbb{R}_+$ of the *radial compactification* $\overline{E}$ of E [69].

We write $N(T) = T^{-1}(0)$ resp. $R(T) = T(V)$ for the *kernel* resp. *range* of a linear operator $T : V \to W$. Finally, we agree to denote the canonical projection of any bundle always with the letter π.

Acknowledgments:

It is a pleasure to thank Richard Melrose for suggesting the study of the 0-calculus; his generous advice has been of invaluable help for the progress of this paper: large parts of the manuscript are influenced by his ideas and insights.

We are greatly indebted to B. Gramsch for drawing our interest to the interplay between functional and pseudodifferential analysis, and for constant support during a long period of time. We like to thank Rafe Mazzeo for his particular interest in this project and many useful and enlightening remarks.

We are grateful to F. Baldus, O. Caps, S. Moroianu, V. Nistor, Th. Schick, E. Schrohe, B.-W. Schulze, and B. Vaillant whom we consulted on several occassions.

We wish to thank the Massachsetts Institute of Technology and the SFB 478 *Geometrische Strukturen in der Mathematik* where parts of the paper were written for the invitation and warm hospitality, and the German Academic Exchange Service (DAAD) for financial support.

Part 1

Fredholm theory for 0-pseudodifferential operators

CHAPTER 1

Review on basic objects of 0-geometry

In this section, we recall the basic geometric objects underlying the 0-pseudo-differential calculus. Throughout this paper, X denotes a smooth, compact, n-dimensional manifold with boundary ∂X. For notational simplicity, let us always assume that ∂X is connected.

1.1. The 0-structure algebra

DEFINITION 1.1.1. The space of all smooth vector fields on X that vanish at the boundary is denoted by $\mathcal{V}_0(X)$. The elements of $\mathcal{V}_0(X)$ are called *0-vector fields*.

Then $\mathcal{V}_0(X)$ is a Lie subalgebra of the Lie algebra $\mathcal{V}(X)$ of all smooth vector fields on X. Moreover, 0-vector fields form a sheaf of $\mathcal{C}^\infty(X)$-modules, and we have $\mathcal{V}_0(X) = \varrho_N \mathcal{V}(X)$ for any defining function $\varrho_N : X \to \overline{\mathbb{R}}_+$ of the boundary of X. To give a local description of 0-vector fields, first note that elements of $\mathcal{V}_0(X)$ are obviously unrestricted in the interior of X.

LEMMA 1.1.2. *Let $q \in \partial X$ be arbitrary, and $(x,y) : X \supseteq V \longrightarrow \overline{\mathbb{R}}_+ \times \mathbb{R}_y^{n-1}$ be local coordinates near q. Then $\mathcal{V}_0(X)|_V$ is spanned over $\mathcal{C}^\infty(V)$ by the vector fields*
$$x\partial_x \text{ and } x\partial_{y_j}, j = 1, \ldots, n-1.$$

As a consequence, we see that $\mathcal{V}_0(X)$ is a finitely generated, projective $\mathcal{C}^\infty(X)$-module; thus, there exists a vector bundle ${}^0TX$ of rank $n = \dim X$ over X together with a natural map $j^0 : {}^0TX \longrightarrow TX$ such that $\mathcal{V}_0(X) = j^0(\mathcal{C}^\infty(X, {}^0TX))$. The bundle ${}^0TX$ is said to be the *0-tangent bundle*, its dual ${}^0T^*X$ the *0-cotangent bundle*. Note that the fibers of ${}^0TX$ are given by
$${}^0T_qX := \mathcal{V}_0(X)/\mathcal{I}_q\mathcal{V}_0(X), \quad q \in \partial X,$$
where $\mathcal{I}_q := \{f \in \mathcal{C}^\infty(X) : f(q) = 0\}$. The natural map $(j^0)_q : {}^0T_qX \to T_qX$ is induced by the natural embedding of $\mathcal{V}_0(X) \hookrightarrow \mathcal{V}(X)$; in particular, $(j^0)_p$ is an isomorphism for $p \notin \partial X$, and $(j^0)_q$ vanishes for $q \in \partial X$.

Let $q \in \partial X$ be arbitrary. Using the identity $[fV, W] = f[V, W] - W(f)V$ and Lemma 1.1.2 we obtain
$$[\mathcal{I}_q\mathcal{V}_0(X), \mathcal{V}_0(X)] \subseteq \mathcal{I}_q\mathcal{V}_0(X),$$
i.e. $\mathcal{I}_q\mathcal{V}_0(X)$ is an ideal in the Lie algebra $\mathcal{V}_0(X)$, thus, ${}^0T_qX$ has a Lie algebra structure as well. For a Lie-algebra $\mathfrak{g}$, let us define the *commutator ideal* of $\mathfrak{g}$ by $[\mathfrak{g}, \mathfrak{g}] := \mathrm{LH}\{[a, b] \in \mathfrak{g} : a, b \in \mathfrak{g}\}$. By iteration, we obtain a sequence of ideals in $\mathfrak{g}$, namely
$$\begin{aligned} \mathfrak{g}^{(0)} &:= \mathfrak{g} \\ \mathfrak{g}^{(k+1)} &:= [\mathfrak{g}^{(k)}, \mathfrak{g}^{(k)}], k \in \mathbb{N}_0. \end{aligned}$$

Recall that the Lie algebra $\mathfrak{g}$ is called *solvable* if there exists a $k \in \mathbb{N}$ such that $\mathfrak{g}^{(k)} = \{0\}$.

Note that there is a natural map of vector bundles ${}^0TX \longrightarrow {}^bTX$. Let us denote the restriction of the kernel $N({}^0TX \longrightarrow {}^bTX)$ of this map to the boundary ∂X by ${}^0T\partial X$.

LEMMA 1.1.3. *Let* $q \in \partial X$. *Then we have* $({}^0T_qX)^{(1)} := [{}^0T_qX, {}^0T_qX] = {}^0T_q\partial X$ *and* $({}^0T_qX)^{(2)} = \{0\}$, *i.e.* ${}^0T_qX$ *is a solvable Lie-algebra.*

Proof: Because of $[fV, gW] \equiv f(q)g(q)[V,W]$ modulo $\mathcal{I}_q\mathcal{V}_0(X)$ for $f, g \in \mathcal{C}^\infty(X)$ and $V, W \in \mathcal{V}_0(X)$, we can restrict ourselves to the basic vector fields of Lemma 1.1.2. The only non-trivial commutator is $[x\partial_x, x\partial_y] = x\partial_y$. This completes the proof. $\square$

In particular, we have a short exact sequence of Lie algebras

$$(1.1) \qquad 0 \longrightarrow {}^0T\partial X \longrightarrow {}^0TX|_{\partial X} \longrightarrow {}^0TX|_{\partial X}/{}^0T\partial X \longrightarrow 0.$$

The choice of a boundary defining function $\varrho_N : X \to \overline{\mathbb{R}}_+$ for X identifies ${}^0T\partial X$ and ${}^0TX|_{\partial X}/{}^0T\partial X$ with some more commonly known objects. First, the map

$$(1.2) \qquad {}^0T\partial X \longrightarrow T\partial X : [W] \longmapsto [(1/\varrho_N W)|_{\partial X}]$$

yields an identification ${}^0T\partial X \to T\partial X$.

LEMMA 1.1.4. *For* $V \in \mathcal{V}_0(X)$, *we have* $V\varrho_N \in \varrho_N \mathcal{C}^\infty(X)$ *for all* $V \in \mathcal{V}_0(X)$, *and*

$$(1.3) \qquad {}^0TX|_{\partial X} \longrightarrow \partial X \times \mathbb{R} : [V] \longmapsto \big(\pi[V], (1/\varrho_N V \varrho_N)|_{\pi[V]}\big)$$

leads to an identification ${}^0TX|_{\partial X}/{}^0T\partial X \to \partial X \times \mathbb{R}$.

1.2. The extended 0-blow up

As already mentioned in the introduction, 0-pseudodifferential operators are defined by characterizing the singularities of their Schwartz kernels on X^2. A relatively simple description of these singularities can be given on a related space, the so-called *0-double space* X_0^2. It is a compact manifold with corners together with a surjective blow-down map $\beta_0^2 : X_0^2 \to X^2$. For reasons that will become clear later on, in this paper we use a slightly different space for the definition of the 0-calculus, and call it the *extended 0-double space* $X_{0,e}^2$. We discuss the relation between the spaces X_0^2 and $X_{0,e}^2$ in Proposition 1.3.1. For the construction of the space $X_{0,e}^2$ we use the general concept of *blowing up p-submanifolds* as developed in detail in the forthcoming book [62]. For a quick review on blow up techniques, see for instance [32, 59, 68, 69].

For the space $X_{0,e}^2$, we need in fact two blow-ups. First, let $B_1 := (\partial X)^2$ be the full corner of X^2, and let

$$(1.4) \qquad \beta_b^2 : X_b^2 := [X^2; B_1] \longrightarrow X^2$$

be the *b-blow up*, see for instance [64] or [68] for more details. Recall that the components $\mathrm{lb} := \overline{(\beta_b^2)^{-1}(\partial X \times (X \setminus \partial X))}^{X_b^2}$ resp. $\mathrm{rb} := \overline{(\beta_b^2)^{-1}((X \setminus \partial X) \times \partial X)}^{X_b^2}$ of the boundary of X_b^2 are called the *left* resp. *right boundary* of the b-double space X_b^2. The lift $\Delta_b := \overline{(\beta_b^2)^{-1}(\Delta \setminus (\partial X)^2)}^{X_b^2}$ of the diagonal $\Delta \subseteq X^2$ is said to be the *b-diagonal*.

1.2. THE EXTENDED 0-BLOW UP

For the second blow-up, let $B_2 := (\beta_b^2)^{-1}(\partial \Delta)$ be the preimage of the boundary of the diagonal $\Delta \subseteq X^2$. The structure of B_2 is in fact rather simple. Note that the composition of the blow-down map β_b^2 with the projection onto one of the factors of X^2 yields a fiber map $\pi : B_2 \to \partial X$.

LEMMA 1.2.1. *There is a canonical diffeomorphism*

$$\Phi : B_2 \longrightarrow [-1, 1] \times \partial X \tag{1.5}$$

of fiber bundles over ∂X.

Proof: Let $\varrho_N : X \to \overline{\mathbb{R}}_+$ be a boundary defining function and ff^b be the boundary hyperface of X_b^2 produced by the b-blow up (1.4). By [**68**, Lemma 4.1], there is a canonical diffeomorphism $\Phi_{\varrho_N} : \mathrm{ff}^b \to [-1,1] \times (\partial X)^2$ such that we have $\beta_b^2|_{\mathrm{ff}^b} = pr_{(\partial X)^2} \circ \Phi_{\varrho_N} : \mathrm{ff}^b \to (\partial X)^2$. Moreover, again by [**68**, Lemma 4.1] we know that the composition of $\Phi_{\varrho_N}|_{B_2}$ with the projection $p : (\partial X)^2 \to \partial X$ onto one of the factors induces a diffeomorphism $\Phi : B_2 \to [-1,1] \times \partial X$ that does not depend on the particular choice of ϱ_N, and makes the following diagram commutative

$$\begin{array}{ccc} B_2 & \xrightarrow{\Phi} & [-1,1] \times \partial X \\ \pi \downarrow & & \downarrow p_2 \\ \partial X & \xrightarrow{id} & \partial X \end{array}$$

$\square$

Blowing up B_2 in X_b^2 gives the *extended 0-blow up*

$$\beta_{0,e}^2 : X_{0,e}^2 := [X_b^2; B_2] \xrightarrow{\beta} X_b^2 \xrightarrow{\beta_b^2} X^2;$$

the manifold $X_{0,e}^2$ is called the *extended 0-double space*. As above, let us denote by $\Delta_{0,e} := \overline{\beta^{-1}(\Delta_b \setminus \partial X_b^2)}^{X_{0,e}^2}$ the *lifted* or *extended 0 diagonal*, and denote by $\mathrm{ff}^{0,e}$ the *extended 0-front face*, i.e. the new boundary hypersurface of $X_{0,e}^2$ produced by the second blow-up. Furthermore, let $T_{0,e}$ resp. $B_{0,e}$ be the lift of the left lb resp. the right rb boundary of the b-double space X_b^2 to $X_{0,e}^2$. We call $T_{0,e}$ resp. $B_{0,e}$ the *top* resp. *bottom face* of $X_{0,e}^2$. Finally, let $F_{0,e}$ be the lift of the b-front face ff^b under the blow-down map β, i.e. $F_{0,e} := \overline{\beta^{-1}(\mathrm{ff}^b \setminus B_2)}^{X_{0,e}^2}$. Note that $F_{0,e}$ has two components if $n = 2$.

Let us consider the structure of the extended 0-front face $\mathrm{ff}^{0,e}$ a little bit more closely. By definition of the blow-up we have

$$\mathrm{ff}^{0,e} = S^+ N B_2 = \left((T^+ X_b^2|_{B_2} / TB_2) \setminus \{0\} \right) / \mathbb{R}_+, \tag{1.6}$$

where $T^+ X_b^2$ stands for the inward pointing part of the tangent bundle. Therefore, $\beta|_{\mathrm{ff}^{0,e}} : \mathrm{ff}^{0,e} \to B_2$ is a fiber bundle whose typical fiber is diffeomorphic to the half-sphere $S_+^{n-1} := \{(\vartheta_0, \vartheta') \in S^{n-1} : \vartheta_0 \geq 0\}$ that is diffeomorphic to the closed unit ball $\overline{D}^{n-1}$ by projection onto the second factor ϑ'.

For local computations, it will be useful to have coordinates near the front face $\mathrm{ff}^{0,e}$. Choose local coordinates

$$(x, y) : X \supseteq U \longrightarrow \overline{\mathbb{R}}_+ \times \mathbb{R}_y^{n-1}$$

near ∂X, and let us denote by (x, y) resp. (x', y') the lift of these coordinates through the projection onto the left resp. the right factor of X^2. The most geometric coordinates near the extended 0-front face ff^{0,e} are certainly polar coordinates $(\tau, \vartheta, \varrho, y') \in [-1,1] \times S_+^{n-1} \times \overline{\mathbb{R}}_+ \times \mathbb{R}_{y'}^{n-1}$ where

$$\tau = \frac{x-x'}{x+x'}, \varrho = \sqrt{(x+x')^2 + |y-y'|^2}, \text{ and } \vartheta = \frac{(x+x', y-y')}{\varrho}. \tag{1.7}$$

In most computations, however, we use the following set of projective coordinates $(\tau, U, r, y') \in [-1,1] \times \mathbb{R}_U^{n-1} \times \overline{\mathbb{R}}_+ \times \mathbb{R}_{y'}^{n-1}$ where

$$\tau = \frac{x-x'}{x+x'}, r = x+x', \text{ and } U = \frac{y-y'}{x+x'} \tag{1.8}$$

near the front face ff^{0,e}. With respect to these coordinates we have

$$\begin{aligned}
\Delta_{0,e} &= \{\tau = 0, \vartheta' = 0\} &= \{\tau = 0, U = 0\}, \\
\text{ff}^{0,e} &= \{\varrho = 0\} &= \{r = 0\}, \\
T_{0,e} &= \{\tau = -1\} &= \{\tau = -1\}, \\
B_{0,e} &= \{\tau = +1\} &= \{\tau = +1\}, \text{ and} \\
F_{0,e} &= \{\vartheta_0 = 0\} &= \{|U| = \infty\},
\end{aligned} \tag{1.9}$$

thus, we see that the projective coordinates are only valid apart from $F_{0,e}$.

PROPOSITION 1.2.2. *The Lie algebra $\mathcal{V}_0(X)$ of 0-vector fields lift, either from the left or from the right factor of X^2, to a Lie subalgebra of $\mathcal{V}_b(X_{0,e}^2)$. The lifted vector fields are transversal to the lifted diagonal. In particular, there are natural isomorphisms*

$$\begin{array}{ccccccc}
{}^0TX & \xrightarrow{\cong} & N\Delta_{0,e} & & N^*\Delta_{0,e} & \xrightarrow{\cong} & {}^0T^*X \\
\downarrow & & \downarrow & \text{and} & \downarrow & & \downarrow \\
X & \xrightarrow{\cong} & \Delta_{0,e} & & \Delta_{0,e} & \xrightarrow{\cong} & X
\end{array}, \tag{1.10}$$

where the diffeomorphism $\Delta_{0,e} \to X$ is induced by the composition of the blow-down map $\beta_{0,e}^2 : X_{0,e}^2 \to X^2$ with the projection onto either the first or the second factor.

Proof: Since the blow-down map $\beta_{0,e}^2 : X_{0,e}^2 \to X^2$ restricted to the interior of $X_{0,e}^2$ is a diffeomorphism, and all constructions are local, it suffices to consider local coordinates. We restrict ourselves to the coordinates (1.8) near the front face. The other cases are similar using appropriate coordinates. A straightforward computation shows that the basic vector fields of Lemma 1.1.2 lift through the projection onto the left factor of X^2 to

$$\begin{aligned}
x\partial_x &\rightsquigarrow \frac{1}{2}(1+\tau)r\partial_r + \frac{1}{2}(1-\tau^2)\partial_\tau - \frac{1}{2}(1+\tau)U\partial_U, \text{ and} \\
x\partial_y &\rightsquigarrow \frac{1}{2}(1+\tau)\partial_U,
\end{aligned}$$

which completes the proof because of (1.9). The lift through the projection onto the right factor follows by symmetry. $\square$

1.3. Relation to the 0-double space X_0^2

Let $\partial\Delta \subseteq X^2$ be the boundary of the diagonal. Recall that in [54, Section 2.B] or [58, Section 3] the 0-*double space* has been defined by

$$\beta_0^2 : X_0^2 := [X^2; \partial\Delta] \longrightarrow X^2.$$

It is a compact manifold with corners up to codimension 3. As for the extended 0-double space, let ff^0 be the new face of X_0^2, and T_0 resp. B_0 be the lift of the left resp. the right boundary of X^2. Let us now compare the 0-double space X_0^2 and the extended 0-double space $X_{0,e}^2$ from a more systematic point of view.

The system $\partial \Delta \subseteq (\partial X)^2 \subseteq X^2$ of submanifolds of the manifolds with corners X^2 is a two-element chain of closed p-submanifolds in the sense of [62], hence $B_2 = (\beta_b^2)^* \partial \Delta$ as well as $B_0 := (\beta_0^2)^*(\partial X)^2$ are closed p-submanifolds of X_b^2 and X_0^2, respectively, and we can consider the blow ups

$$X_{0,e}^2 := [X_b^2; B_2] =: [X^2; (\partial X)^2; \partial \Delta] \text{ resp. } [X_0^2; B_0] =: [X^2; \partial \Delta; (\partial X)^2]$$

of X_b^2 resp. X_0^2 along B_2 resp. B_0.

PROPOSITION 1.3.1. *The identity on $X^2 \setminus (\partial X)^2$ induces a natural diffeomorphism $\Phi : [X^2; (\partial X)^2; \partial \Delta] \to [X^2; \partial \Delta; (\partial X)^2]$ making the following diagram commutative:*

(1.11)
$$\begin{array}{ccc}
[X^2; (\partial X)^2; \partial \Delta] & \stackrel{\Phi}{\longrightarrow} & [X^2; \partial \Delta; (\partial X)^2] \\
\beta \downarrow & & \downarrow \\
[X^2; (\partial X)^2] & \# & [X^2; \partial \Delta] \\
\beta_b^2 \downarrow & & \downarrow \beta_0^2 \\
X^2 & \stackrel{\mathrm{id}}{\longrightarrow} & X^2
\end{array}$$

Here, the vertical maps are the natural blow-down maps.

For a proof, we refer to [62, Proposition 5.8.1] – see also [69, Section 18].

The main technical advantage of the extended 0-double space is that the variable $\tau = \frac{1+\tau}{1-\tau}$ becomes a global variable near the extended 0-front face $\mathrm{ff}^{0,e}$ – see Lemma 1.2.1.

1.4. The extended 0-triple space $X_{0,e}^3$

To understand the composition of 0-pseudodifferential operators it is convenient to have also an extended "triple" version of the extended 0-double space $X_{0,e}^2$. For similar constructions, we refer to [44, 55, 60, 69]. A triple version of the 0-double space can be found in [65] or as a special case of the edge triple space in [55]. Here, we are going to construct an *extended version* of this triple space corresponding to the extended 0-double space. We need the following projections:

$$\begin{aligned}
\pi_C &: X^3 \longrightarrow X^2 : (q_1, q_2, q_3) \longmapsto (q_1, q_3), \\
\pi_F &: X^3 \longrightarrow X^2 : (q_1, q_2, q_3) \longmapsto (q_2, q_3), \text{ and} \\
\pi_S &: X^3 \longrightarrow X^2 : (q_1, q_2, q_3) \longmapsto (q_1, q_2).
\end{aligned}$$

For $O \in \{C, F, S\}$, let us denote by $B_{1,O} := \pi_O^{-1}(B_1) \subseteq X^3$ the lift of the corner $B_1 = (\partial X)^2 \subseteq X^2$ under the projection π_O, and $B_{1,T} := (\partial X)^3$ the *full corner* of X^3. It is straightforward to check that the spaces $B_{1,O}$, $O = C, F, S, T$ are p-submanifolds of the manifold with corners X^3 satisfying

$$B_{1,T} = B_{1,O} \cap B_{1,O'} \text{ for all } O, O' \in \{C, F, S\} \text{ with } O \neq O'.$$

Let $\beta_b^3 : X_b^3 := [X^3; B_{1,T}; B_{1,C}; B_{1,F}; B_{1,S}] \to X^3$ be the composition of the corresponding blow-down maps. Note that X_b^3 is exactly the *b-triple space* as constructed in [**65**], [**73**, Appendix], or a special case of the edge triple space in [**55**].

Similarly, let $B_{2,O} := \pi_O^{-1}(B_2) \subseteq X^3$ be the lift of the boundary of the diagonal $\partial \Delta \subseteq X^2$ under π_O. Because of $B_{2,O} \subseteq B_{1,O}$, the p-submanifolds $B_{2,O}$ lift to pairwise disjoint p-submanifolds $B'_{2,O} := (\beta_b^3)^* B_{2,O}$ in X_b^3. Let $B'_{1,T} := (\beta_b^3)^* B_{1,T}$ be the lift of the full-corner under β_b^3, and $B_O := B'_{2,O} \cap B'_{1,T} \subseteq X_b^3$, $O \in \{C, F, S\}$. Then B_O, $O \in \{C, F, S\}$, is a system of pairwise disjoint p-submanifolds in X_b^3, and we have

(1.12) $$B_O \cap B'_{2,O'} = \emptyset \text{ for all } O, O' \in \{C, F, S\} \text{ with } O \neq O'.$$

We can now define the *extended 0-triple space* $X_{0,e}^3$ by

(1.13) $$\beta_{0,e}^3 : X_{0,e}^3 := [X_b^3; B_C; B_F; B_S; B'_{2,C}; B'_{2,F}; B'_{2,S}] \longrightarrow X^3,$$

where $\beta_{0,e}^3$ is the composition of the obvious blow down map to X_b^3 with the b-triple blow down map $\beta_b^3 : X_b^3 \to X^3$.

The main properties of the extended 0-triple space $X_{0,e}^3$ are collected in the following proposition.

PROPOSITION 1.4.1. *The extended 0-triple space $X_{0,e}^3$ is a compact manifold with corners such that:*

(a) *The action of the symmetry group in three variables acting on the three factors of X^3 lifts to diffeomorphisms of $X_{0,e}^3$.*

(b) *There exist b-fibrations $\pi_O^{0,e} : X_{0,e}^3 \longrightarrow X_{0,e}^2$, $O = C, F, S$ making the following diagram commutative:*

(1.14) $$\begin{array}{ccc} X_{0,e}^3 & \xrightarrow{\pi_O^{0,e}} & X_{0,e}^2 \\ \beta_{0,e}^3 \downarrow & & \downarrow \beta_{0,e}^2 \\ X^3 & \xrightarrow{\pi_O} & X^2 \end{array}.$$

(c) *The extended 0-diagonal $\Delta_{0,e}$ in $X_{0,e}^2$ lifts under $\pi_O^{0,e}$ to a p-submanifold $\Delta_O^{0,e} \subseteq X_{0,e}^3$ such that*

$$\pi_{O'}^{0,e}|_{\Delta_O^{0,e}} : \Delta_O^{0,e} \longrightarrow X_{0,e}^2$$

is a diffeomorphism for all $O, O' \in \{C, F, S\}$ with $O \neq O'$.

Proof: In order to avoid clumsy notation, we do not introduce new notations for the lifts of p-submanifolds under several blow up maps, but instead indicate, if necessary, in which space these submanifolds have to be considered.

The symmetry condition (a) follows immediately from the symmetry of the construction of the space $X_{0,e}^3$. By this symmetry of $X_{0,e}^3$, it suffices to construct the map $\pi_F^{0,e} : X_{0,e}^3 \longrightarrow X^3$ and to check (b) and (c). We define $\pi_F^{0,e}$ as the composition of the following maps:

$$
\begin{align}
(1.15)\quad \pi_F^{0,e} : X_{0,e}^3 &= [X_b^3; B_F; B_C; B_S; B'_{2,F}; B'_{2,C}; B'_{2,S}] \\
(1.16)\quad &\longrightarrow [X_b^3; B_F; B_C; B_S; B'_{2,F}] \\
(1.17)\quad &\cong [X^3; B_{1,T}; B_{1,F}; B_{1,C}; B_{1,S}; B_F; B'_{2,F}; B_C; B_S] \\
(1.18)\quad &\longrightarrow [X^3; B_{1,T}; B_{1,F}; B_{1,C}; B_{1,S}; B_F; B'_{2,F}] \\
(1.19)\quad &\cong [X^3; B_{1,T}; B_{1,F}; B_F; B'_{2,F}; B_{1,C}; B_{1,S}] \\
(1.20)\quad &\longrightarrow [X^3; B_{1,T}; B_{1,F}; B_F; B'_{2,F}] \\
(1.21)\quad &\cong [X^3; B_{1,F}; B_F; B_{1,T}; B'_{2,F}] \\
(1.22)\quad &\cong [X^3; B_{1,F}; B_F; B'_{2,F}; B_{1,T}] \\
(1.23)\quad &\longrightarrow [X^3; B_{1,F}; B_F; B'_{2,F}] \\
(1.24)\quad &\cong [X \times X^2; X \times B_1; X \times B_2] \\
(1.25)\quad &\cong X \times X_{0,e}^2 \\
(1.26)\quad &\longrightarrow X_0^2,
\end{align}
$$

where (1.15) is the definition of the extended 0-triple space, (1.16), (1.18), (1.20), and (1.23) are the obvious blow-down maps, (1.26) is the projection onto the second factor, and (1.25) is induced by the definition of the extended 0-double space $X_{0,e}^2$.

For the diffeomorphism (1.17) we use the definition of the b-triple space X_b^3 and $B'_{2,F} \cap B_C = B'_{2,F} \cap B_S = \emptyset$ by (1.12). The diffeomorphism (1.19) follows from $B_F \cap B_{1,O} = B'_{2,F} \cap B_{1,O} = \emptyset$ in $[X^3; B_{1,T}]$ for $O = C, S$. For (1.21) we apply the commutativity of the blow up [**62**, Proposition 5.8.1] because of $B_{1,T} \subseteq B_{1,F}$ in X^3, and $B_F \subseteq B_{1,T}$ in $[X^3; B_{1,F}]$. Since $B_{1,T}$ and $B'_{2,F}$ are disjoint in the space $[X^3; B_{1,F}; B_F]$ where its intersection B_F is blown up, the diffeomorphism (1.22) follows. Note that $B_{1,T}$ and $B'_{2,F}$ are transversal p-submanifolds of the hypersurface $X \times \mathrm{ff}^b$ in $X \times X_b^2 \cong [X^3; B_{1,F}]$.

Since in both possible ways from $X_{0,e}^3$ to X^2 in (1.14) finally all blow ups are blown down again, and the order does not matter by the arguments above, the commutativity of (1.14) follows.

As in the proof of [**69**, Lemma 16] we see using [**69**, Proposition 16] that $\pi_F^{0,e}$ is a b-fibration, and (c) follows by continuity from the interior. $\square$

1.5. 0-densities

The last bit of basic 0-geometry are densities adapted to the calculus. Apply the smooth functor Ω^α of α-densities to the 0-tangent bundle ^0TX to obtain a smooth vector bundle, denoted $^0\Omega^\alpha(X)$, the bundle of smooth 0-α-*densities*. The choice of local coordinates (x, y) trivializes the bundle $^0\Omega^\alpha(X)|_U$. A non vanishing section is given by $\left|\frac{dx}{x^n}dy\right|^\alpha$. In particular, we obtain a well-defined integral

$$\int_X : \dot{\mathcal{C}}^\infty(X, {}^0\Omega^1(X)) \longrightarrow \mathbb{C}$$

that does not extend naturally to $\mathcal{C}^\infty(X, {}^0\Omega^1)$ because of the singular factor x^{-n} in the 0-density near the boundary. However, it leads to a natural scalar product on $\dot{\mathcal{C}}^\infty(X, {}^0\Omega^{1/2})$:

(1.27) $\qquad <f,g>_{L^2(X,{}^0\Omega^{1/2})} := \int_X f\bar{g}$ for $f,g \in \dot{\mathcal{C}}^\infty(X, {}^0\Omega^{1/2})$.

On the extended 0-double space $X^2_{0,e}$, for technical reasons we mainly use the *extended 0-kernel half-density bundle* $KD^{1/2}_{0,e} := \varrho_{\mathrm{ff}^{0,e}}^{-n/2} \Omega^{1/2}(X^2_{0,e})$. It is completely characterized by the space of its $\mathcal{C}^\infty$-sections [**68**, Lemma 8.6], namely $\mathcal{C}^\infty(X^2_{0,e}, KD^{1/2}_{0,e}) = \varrho_{\mathrm{ff}^{0,e}}^{-n/2} \mathcal{C}^\infty(X^2_{0,e}, \Omega^{1/2})$.

REMARK 1.5.1. The blow-down map $\beta^2_{0,e}: X^2_{0,e} \to X^2$ induces, via pull-back and duality, isomorphisms

(1.28) $(\beta^2_{0,e})^* : \dot{\mathcal{C}}^\infty(X^2, {}^0\Omega^{1/2} \boxtimes {}^0\Omega^{1/2}) \xrightarrow{\cong} \dot{\mathcal{C}}^\infty(X^2_{0,e}, KD^{1/2}_{0,e})$, and

(1.29) $(\beta^2_{0,e})_* : \quad \mathcal{C}^{-\infty}(X^2_{0,e}, KD^{1/2}_{0,e}) \xrightarrow{\cong} \mathcal{C}^{-\infty}(X^2, {}^0\Omega^{1/2} \boxtimes {}^0\Omega^{1/2})$.

However, because of

$$\mathcal{C}^\infty(X^2, {}^0\Omega^{1/2} \boxtimes {}^0\Omega^{1/2}) \subsetneq (\beta^2_{0,e})_*((\varrho_{T_{0,e}} \varrho_{B_{0,e}})^{-n/2} \varrho_{F_{0,e}}^{(1-2n)/2} \mathcal{C}^\infty(X^2_{0,e}, KD^{1/2}_{0,e}))$$

there are "more" smooth half-densities on $X^2_{0,e}$ than on X^2. This is the flexibility needed for a simple description of the Schwartz kernels of 0-pseudodifferential operators.

CHAPTER 2

The small 0-calculus and the 0-calculus with bounds

2.1. The Schwartz kernel theorem revisited

Since 0-pseudodifferential operators are defined by the singularities of their Schwartz kernels, let us first recall the following version of the Schwartz kernel theorem.

PROPOSITION 2.1.1. *There is a one-to-one correspondence between the continuous linear operators*

$$\dot{\mathcal{C}}^\infty(X, {}^0\Omega^{1/2}) \longrightarrow \mathcal{C}^{-\infty}(X, {}^0\Omega^{1/2})$$

on the one hand, and the space of extendible *distributions* $\mathcal{C}^{-\infty}(X_{0,e}^2, KD_{0,e}^{1/2})$ *on the other hand.*

If $\kappa_A \in \mathcal{C}^{-\infty}(X_{0,e}^2, KD_{0,e}^{1/2})$ denotes the lifted Schwartz kernel of a bounded linear operator $A : \dot{\mathcal{C}}^\infty(X, {}^0\Omega^{1/2}) \to \mathcal{C}^{-\infty}(X, {}^0\Omega^{1/2})$ under this correspondence, then we have for all $f \in \dot{\mathcal{C}}^\infty(X, {}^0\Omega^{1/2})$ and $\nu \in \mathcal{C}^\infty(X, {}^0\Omega^{1/2})$

$$(2.1) \qquad \nu A f \;=\; (\pi_{L;0,e})_* \left(\pi_{R;0,e}^*(f) \pi_{L;0,e}^*(\nu) \kappa_A \right)$$

$$(2.2) \qquad \qquad =\; (\pi_L)_* \left(\pi_R^*(f) \pi_L^*(\nu) (\beta_{0,e}^2)_*(\kappa_A) \right) \in \mathcal{C}^{-\infty}(X, {}^0\Omega^1),$$

where $\pi_{L;0,e} := \pi_L \circ \beta_{0,e}^2 : X_{0,e}^2 \to X$ resp. $\pi_{R;0,e} := \pi_R \circ \beta_{0,e}^2 : X_{0,e}^2 \to X$ are the projections of the extended 0-double space to the left resp. right factor.

Proof: This follows from the isomorphisms (1.28) and (1.29) and a chase through the definitions from the usual Schwartz-kernel theorem on manifolds with corners [68]. □

2.2. The small 0-calculus

DEFINITION 2.2.1. A bounded linear operator

$$A : \dot{\mathcal{C}}^\infty(X, {}^0\Omega^{1/2}) \longrightarrow \mathcal{C}^{-\infty}(X, {}^0\Omega^{1/2})$$

is said to be a *classical 0-pseudodifferential operator of order $m \in \mathbb{C}$* in the *small calculus* provided

$$\kappa_A \in \left\{ \kappa \in I_{cl}^m(X_{0,e}^2, \Delta_{0,e}; KD_{0,e}^{1/2}) : \kappa \equiv 0 \text{ at } \partial X_{0,e}^2 \setminus \mathrm{ff}^{0,e} \right\} \subseteq \mathcal{C}^{-\infty}(X_{0,e}^2, KD_{0,e}^{1/2}),$$

where $\equiv$ means vanishing to infinite order. The space of all classical 0-pseudodifferential operators of order $m \in \mathbb{C}$ is denoted by $\Psi_0^m(X; {}^0\Omega^{1/2})$.

For the definition and basic properties of the spaces $I_{cl}^m(X, Y; E)$ of (classical) conormal distributions we refer to [35], [36, Section 18.2], or [97]. Let us mention that we could also have defined 0-pseudodifferential operators using the space

$I^m(X^2_{0,e}, \Delta_{0,e}; KD^{1/2}_{0,e})$, $m \in \mathbb{R}$, of conormal distributions instead of the space of classical one's. In fact, most of the results that follow are true for this slightly larger class, too. However, since we prefer the principal symbol to be a function we restricted ourselves to the classical case, and do not even mention this point in the notation. Note that the 0-kernel half-density bundle is chosen in such a way that the lifted Schwartz kernel of the identity corresponds to a delta-section of the lifted diagonal $\Delta_{0,e}$ that is smooth up to the extended 0-front face $\mathrm{ff}^{0,e}$.

REMARK 2.2.2. (a) By Proposition 1.3.1, we see that the obvious blow-down map $\beta : X^2_{0,e} \cong [X^2; \partial \Delta; (\partial X)^2] \to X^2_0$ in (1.11) induces an isomorphism $(\beta)_*$ between $\Psi^m_0(X; {}^0\Omega^{1/2})$ and the space

$$\left\{ \kappa \in I^m_{cl}(X^2_0, \Delta_0; KD^{1/2}_0) : \kappa \equiv 0 \text{ at } \partial X^2_0 \setminus \mathrm{ff}^0 \right\};$$

here the 0-kernel half-density bundle is determined by its space of smooth sections $\mathcal{C}^\infty(X^2_0, KD^{1/2}_0) = \varrho^{-n/2}_{\mathrm{ff}^0} \mathcal{C}^\infty(X^2_0, \Omega^{1/2})$; therefore, the small 0-calculus defined in Definition 2.2.1 coincides with the small 0-calculus defined in [54] or [58] using the 0-double space X^2_0.

(b) As in the case of classical pseudodifferential operators on closed manifolds the symbol topology near the lifted diagonal $\Delta_{0,e}$, and the $\mathcal{C}^\infty$-topology on $X^2_{0,e} \setminus \Delta_{0,e}$ endows the spaces $\Psi^m_0(X; {}^0\Omega^{1/2})$ with natural Fréchet-topologies.

2.3. Basic properties of the small 0-calculus

We collect some immediate consequences from the definition of the small calculus. Most of the results are well-known because of Remark 2.2.2(a), thus, we only sketch the proofs. More elaborated proofs can be found in [53, 54, 58]. Let us start with the homogeneous principal symbol. Recall the symbol map for conormal distributions; it is given by first taking the normal bundle $N\Delta_{0,e}$ of $\Delta_{0,e}$ in $X^2_{0,e}$ as a local model for $X^2_{0,e}$ near $\Delta_{0,e}$, and then applying the invariant Fourier transform in the fiber direction – for more details see [35, 36, 97]. Thus, we get a map

$$(2.3) \quad \sigma : I^m_{cl}(X^2_{0,e}, \Delta_{0,e}; KD^{1/2}_{0,e}) \longrightarrow S^{[m]}(N^*\Delta_{0,e}) \otimes \Omega_{\mathrm{fiber}}(N^*\Delta_{0,e}) \otimes KD^{1/2}_{0,e},$$

where $\Omega_{\mathrm{fiber}}(E)$ stands for the *fiber density bundle* of a vector bundle E. The density factor is in fact canonically trivial: indeed, using the identification (1.10) we obtain

$$\begin{aligned}
\Omega_{\mathrm{fiber}}(N^*\Delta_{0,e}) \otimes KD^{1/2}_{0,e} &\cong \Omega_{\mathrm{fiber}}({}^0T^*X) \otimes \varrho^{-n/2}_{\mathrm{ff}^{0,e}} \Omega^{1/2}(N\Delta_{0,e}) \\
&\cong \Omega_{\mathrm{fiber}}({}^0T^*X) \otimes \varrho^{-n/2}_{\partial X} \Omega^{1/2}(X) \otimes \Omega^{1/2}_{\mathrm{fiber}}({}^0TX) \\
&\cong \Omega^{-1}_{\mathrm{fiber}}({}^0TX) \otimes \Omega^{1/2}_{\mathrm{fiber}}({}^0TX) \otimes \Omega^{1/2}_{\mathrm{fiber}}({}^0TX) \\
&\cong \mathbb{C}.
\end{aligned}$$

Removing the density factor in (2.3) finally leads to the following short exact sequence

$$(2.4) \quad 0 \longrightarrow \Psi^{m-1}_0(X; {}^0\Omega^{1/2}) \longrightarrow \Psi^m_0(X; {}^0\Omega^{1/2}) \xrightarrow{{}^0\sigma^{(m)}} S^{[m]}({}^0T^*X) \longrightarrow 0.$$

The map ${}^0\sigma^{(m)} : \Psi^m_0(X; {}^0\Omega^{1/2}) \to S^{[m]}({}^0T^*X)$ is called the *homogeneous principal symbol*. By choosing a defining function of the "sphere at infinity", the *0-cosphere* bundle $^0S^*X := (^0T^*X \setminus \{0\})/\mathbb{R}_+$ in the *radial compactification* $^0\overline{T}^*X$ of the 0-cotangent bundle $^0T^*X$ [**69**, Section 1] we can identify $S^{[m]}(^0T^*X)$ with the space $\mathcal{C}^\infty(^0S^*X)$.

Let us now consider $\mathcal{C}^\infty$-mapping properties of operators in the small 0-calculus. If $A \in \Psi_0^m(X; {}^0\Omega^{1/2})$, $f \in \dot{\mathcal{C}}^\infty(X, {}^0\Omega^{1/2})$ and $\nu \in \mathcal{C}^\infty(X, {}^0\Omega^{1/2})$ are arbitrary, then $\pi^*_{R;0,e}(f)\pi^*_{L;0,e}(\nu)\kappa_A$ is a distribution on $X_{0,e}^2$ that is conormal to $\Delta_{0,e}$, and vanishes to infinite order at all boundary hyperfaces of $X_{0,e}^2$, thus, we obtain by (2.1), and the push-forward theorem for conormal distributions [**19**, Proposition B7.11]

$$\nu A f = (\pi_{L;0,e})_* \left(\pi^*_{R;0,e}(f)\pi^*_{L;0,e}(\nu)\kappa_A\right) \in \dot{\mathcal{C}}^\infty(X, {}^0\Omega^1).$$

Since $\nu \in \mathcal{C}^\infty(X, {}^0\Omega^{1/2})$ was arbitrary, we have $Af \in \dot{\mathcal{C}}^\infty(X, {}^0\Omega^{1/2})$. Therefore, by the closed graph theorem, each $A \in \Psi_0^m(X; {}^0\Omega^{1/2})$ induces a bounded, linear operator $A : \dot{\mathcal{C}}^\infty(X, {}^0\Omega^{1/2}) \to \dot{\mathcal{C}}^\infty(X, {}^0\Omega^{1/2})$; in particular, composition of operators in the small calculus is well-defined.

PROPOSITION 2.3.1. *For $m_1, m_2 \in \mathbb{R}$, we have*

$$\Psi_0^{m_1}(X; {}^0\Omega^{1/2}) \circ \Psi_0^{m_2}(X; {}^0\Omega^{1/2}) \subseteq \Psi_0^{m_1+m_2}(X; {}^0\Omega^{1/2}).$$

Moreover, for $A_j \in \Psi_0^{m_j}(X; {}^0\Omega^{1/2})$, we have

(2.5) $\quad {}^0\sigma^{(m_1+m_2)}(A_1 A_2) = {}^0\sigma^{(m_1)}(A_1) {}^0\sigma^{(m_2)}(A_2) \in S^{[m_1+m_2]}(^0T^*X).$

Proof: The proof is appropriately adapted from the proof of the corresponding composition formula for Θ-pseudodifferential operators in [**19**, Theorem 11.30]. Choose a smooth, non-vanishing section $\mu \in \mathcal{C}^\infty(X_{0,e}^2, KD_{0,e}^{1/2})$, thus, every section $\kappa \in \mathcal{C}^{-\infty}(X_{0,e}^2, KD_{0,e}^{1/2})$ can be written as $\kappa = \widehat{\kappa} \cdot \mu$ with $\widehat{\kappa} \in \mathcal{C}^{-\infty}(X_{0,e}^2)$. Then the lifted Schwartz kernel $\kappa_{A_1 A_2}$ of $A_1 A_2$ satisfies
(2.6)
$$\widehat{\kappa}_{A_1 A_2} \cdot \mu^2 = (\pi_C^{0,e})_* \left[(\pi_S^{0,e})^* \widehat{\kappa}_{A_1} \cdot (\pi_F^{0,e})^* \widehat{\kappa}_{A_2} \cdot (\pi_S^{0,e})^*\mu \otimes (\pi_F^{0,e})^*\mu \otimes (\pi_C^{0,e})^*\mu \right],$$

i.e. it is the push-forward of a distributional density on $X_{0,e}^3$. An application of [**19**, Proposition B7.20] gives

$$\widehat{\kappa}_{A_1 A_2} \cdot \mu^2 \in \left\{ \kappa \in I_{cl}^m(X_{0,e}^2, \Delta_{0,e}; KD_{0,e}^1) : \kappa \equiv 0 \text{ at } \partial X_{0,e}^2 \setminus \mathrm{ff}^{0,e} \right\},$$

which implies $A_1 A_2 \in \Psi_0^{m_1+m_2}(X; {}^0\Omega^{1/2})$ as well as the identity (2.5) for the homogeneous principal symbol.

For applying [**19**, Proposition B7.20] we first have to transform the density $(\pi_S^{0,e})^*\mu \otimes (\pi_F^{0,e})^*\mu \otimes (\pi_C^{0,e})^*\mu$ into a b-density on $X_{0,e}^3$. This transformation leads to additional powers of defining functions of faces of $X_{0,e}^3$. However, these factors can be ignored because of the rapid vanishing of the kernels at the faces other than the front face. The fact that the density factors fit together follows by taking $A_1 = id$ in (2.6). $\square$

PROPOSITION 2.3.2. *Let $A \in \Psi_0^m(X; {}^0\Omega^{1/2})$ be arbitrary. Then there exists a (uniquely determined) 0-pseudodifferential operator $A^\sharp \in \Psi_0^m(X; {}^0\Omega^{1/2})$ such that*

(2.7) $\quad <Af, g>_{L^2(X, {}^0\Omega^{1/2})} = <f, A^\sharp g>_{L^2(X, {}^0\Omega^{1/2})}$ *for all $f, g \in \dot{\mathcal{C}}^\infty(X, {}^0\Omega^{1/2})$.*

Moreover, we have ${}^0\sigma^{(m)}(A^\sharp) = \overline{{}^0\sigma^{(m)}(A)}$. The operator $A^\sharp$ is called the formal adjoint *of A.*

Proof: Let $c : \mathcal{C}^{-\infty}(X^2_{0,e}, KD^{1/2}_{0,e}) \to \mathcal{C}^{-\infty}(X^2_{0,e}, KD^{1/2}_{0,e})$ be the operator of complex conjugation, and let $\alpha : X^2 \longrightarrow X^2$ be the *flip*. Then α extends to a diffeomorphism $\alpha : X^2_{0,e} \to X^2_{0,e}$, and the operator with Schwartz kernel $\kappa := c\alpha^*(\kappa_A)$ satisfies (2.7). On the other hand, $\Psi^m_0(X; {}^0\Omega^{1/2})$ is clearly invariant under c and α^* which gives $A^\sharp \in \Psi^m_0(X; {}^0\Omega^{1/2})$. The property of the homogeneous principal symbol is then an immediate consequence of the definition of the symbol map for conormal distributions in (2.3). $\square$

Therefore, we can extend the action of $A \in \Psi^m_0(X; {}^0\Omega^{1/2})$ to $\mathcal{C}^{-\infty}(X, {}^0\Omega^{1/2})$ by defining

$$A : \mathcal{C}^{-\infty}(X, {}^0\Omega^{1/2}) \longrightarrow \mathcal{C}^{-\infty}(X, {}^0\Omega^{1/2}) : u \longmapsto \left[\dot{\mathcal{C}}^\infty(X, {}^0\Omega^{1/2}) \ni \varphi \longmapsto u\left(A^\sharp(\varphi)\right) \right],$$

and as above, we see that we have $A(\mathcal{C}^\infty(X, {}^0\Omega^{1/2})) \subseteq \mathcal{C}^\infty(X, {}^0\Omega^{1/2})$, i.e. the linear operator $A : \mathcal{C}^\infty(X, {}^0\Omega^{1/2}) \to \mathcal{C}^\infty(X, {}^0\Omega^{1/2})$ is continuous by the closed graph theorem. Indeed, it suffices to notice that for $f, \nu \in \mathcal{C}^\infty(X, {}^0\Omega^{1/2})$, the distributional density $\pi^*_{R;0,e}(f)\pi^*_{L;0,e}(\nu)\kappa_A$ is conormal to $\Delta_{0,e}$, and vanishes at all boundary faces to infinite order, except for the extended 0-front face $\mathrm{ff}^{0,e}$, and then to apply again the push-forward theorem.

Let now $\varrho_N : X \to \overline{\mathbb{R}}_+$ be a boundary defining function, and $k \in \mathbb{C}$ be arbitrary. Then multiplication with the function ϱ_N^k induces a continuous linear operator $\varrho_N^k : \dot{\mathcal{C}}^\infty(X, {}^0\Omega^{1/2}) \to \dot{\mathcal{C}}^\infty(X, {}^0\Omega^{1/2})$.

LEMMA 2.3.3. *For $z \in \mathbb{C}$, we have $\varrho_N^{-z} \Psi^m_0(X; {}^0\Omega^{1/2}) \varrho_N^z = \Psi^m_0(X; {}^0\Omega^{1/2})$, and*

$$(2.8) \qquad {}^0\sigma^{(m)}(\varrho_N^{-z} A \varrho_N^z) = {}^0\sigma^{(m)}(A)$$

holds for all $A \in \Psi^m_0(X; {}^0\Omega^{1/2})$. Moreover, for any $A \in \Psi^m_0(X; {}^0\Omega^{1/2})$, the map

$$(2.9) \qquad \mathbb{C} \longrightarrow \Psi^m_0(X; {}^0\Omega^{1/2}) : z \longmapsto \varrho_N^{-z} A \varrho_N^z$$

is analytic.

Proof: It is sufficient to note that the lifted Schwartz kernel κ_B of $B := \varrho_N^{-z} A \varrho_N^z$ is given up to smooth non-vanishing factors by $\kappa_B = \varrho_{B_{0,e}}^z \varrho_{T_{0,e}}^{-z} \kappa_A$. $\square$

DEFINITION 2.3.4. *For $k, m \in \mathbb{C}$, let*

$$\Psi^{m,k}_0(X; {}^0\Omega^{1/2}) := \varrho_N^{-k} \Psi^m_0(X; {}^0\Omega^{1/2}) = \Psi^m_0(X; {}^0\Omega^{1/2}) \varrho_N^{-k}.$$

Clearly, the spaces $\Psi^{m,k}_0(X; {}^0\Omega^{1/2})$ do not depend on the particular choice of the boundary defining function ϱ_N. Furthermore, note that we have

$$(2.10) \quad \Psi^{m,k}_0(X; {}^0\Omega^{1/2}) \subseteq \Psi^{m',k'}_0(X; {}^0\Omega^{1/2}) \iff m' - m \in \mathbb{N}_0 \text{ and } k' - k \in \mathbb{N}_0,$$

which was the reason for defining $\Psi^{m,k}_0(X; {}^0\Omega^{1/2})$ using ϱ_N^{-k} with the negative power. A combination of Proposition 2.3.1 and Lemma 2.3.3 gives

$$(2.11) \qquad \Psi^{m_1,k_1}_0(X; {}^0\Omega^{1/2}) \circ \Psi^{m_2,k_2}_0(X; {}^0\Omega^{1/2}) \subseteq \Psi^{m_1+m_2,k_1+k_2}_0(X; {}^0\Omega^{1/2}),$$

and for $A = B\varrho_N^{-k} \in \Psi^{m,k}_0(X; {}^0\Omega^{1/2})$ the operator $A^\sharp := \varrho_N^{-k} B^\sharp \in \Psi^{m,k}_0(X; {}^0\Omega^{1/2})$ satisfies

$$< Af, g >_{L^2(X, {}^0\Omega^{1/2})} = < f, A^\sharp g >_{L^2(X, {}^0\Omega^{1/2})} \text{ for all } f, g \in \dot{\mathcal{C}}^\infty(X, {}^0\Omega^{1/2}).$$

By (2.8), the choice of boundary defining function $\varrho_N : X \to \overline{\mathbb{R}}_+$, leads to a system of maps

$$^0\sigma^{(m,k)} : \Psi_0^{m,k}(X; {}^0\Omega^{1/2}) \longrightarrow S^{[m]}({}^0T^*X) \cong \mathcal{C}^\infty({}^0S^*X) :$$
$$A \longmapsto {}^0\sigma^{(m)}(\varrho_N^k A) = {}^0\sigma^{(m)}(A\varrho_N^k)$$

that is multiplicative in the obvious sense and satisfies ${}^0\sigma^{(m,k)}(A^\sharp) = \overline{{}^0\sigma^{(m,k)}(A)}$.

REMARK 2.3.5. Since the lift of the defining function ϱ_N through the left resp. the right factor is given by $\varrho_{T_{0,e}}\varrho_{\mathrm{ff}^{0,e}}$ resp. $\varrho_{B_{0,e}}\varrho_{\mathrm{ff}^{0,e}}$, the form of the Schwartz kernel theorem in Proposition 2.1.1 yields the following identification
(2.12)
$$\Psi_0^{m,k}(X; {}^0\Omega^{1/2}) \xleftrightarrow{1\text{-}1} \underbrace{\varrho_{\mathrm{ff}^{0,e}}^{-k} \left\{ \kappa \in I_{cl}^m(X_{0,e}^2, \Delta_{0,e}; KD_{0,e}^{1/2}) : \kappa \equiv 0 \text{ at } \partial X_{0,e}^2 \setminus \mathrm{ff}^{0,e} \right\}}_{\subseteq \mathcal{C}^{-\infty}(X_{0,e}^2, KD_{0,e}^{1/2})}.$$

Asymptotic completeness of the 0-calculus can be proved exactly as asymptotic completeness in the classical case.

PROPOSITION 2.3.6. Let $A_j \in \Psi_0^{m-j,k}(X; {}^0\Omega^{1/2})$, $B_j \in \Psi_0^{m,k-j}(X; {}^0\Omega^{1/2})$, and $C_j \in \Psi_0^{m-j,k-j}(X; {}^0\Omega^{1/2})$, $j \in \mathbb{N}_0$ be arbitrary. Then there exist 0-operators $A, B, C \in \Psi_0^{m,k}(X; {}^0\Omega^{1/2})$ such that we have $A - \sum_{j=0}^{\ell-1} A_j \in \Psi_0^{m-\ell,k}(X; {}^0\Omega^{1/2})$, $B - \sum_{j=0}^{\ell-1} B_j \in \Psi_0^{m,k-\ell}(X; {}^0\Omega^{1/2})$ and $C - \sum_{j=0}^{\ell-1} C_j \in \Psi_0^{m-\ell,k-\ell}(X; {}^0\Omega^{1/2})$. The operators A, B, and C are uniquely determined up to the spaces $\Psi_0^{-\infty,k}(X; {}^0\Omega^{1/2})$, $\Psi_0^{m,-\infty}(X; {}^0\Omega^{1/2})$, and $\Psi_0^{-\infty,-\infty}(X; {}^0\Omega^{1/2})$, respectively. In that case, we write $A \sim \sum_{j=0}^\infty A_j$, $B \sim \sum_{j=0}^\infty B_j$, and $C \sim \sum_{j=0}^\infty C_j$.

By the exactness of (2.4), an operator whose homogeneous principal symbol is invertible, is invertible up to operators of lower order. Using the asymptotic completeness of Proposition 2.3.6 and a formal Neumann series argument, we obtain as usual the invertibility up to operators of order $-\infty$.

PROPOSITION 2.3.7. Let $A \in \Psi_0^{m,k}(X; {}^0\Omega^{1/2})$ be with ${}^0\sigma^{(m,k)}(A)(\zeta) \neq 0$ for all $\zeta \in {}^0T^*X \setminus \{0\}$. Then there exists a $B \in \Psi_0^{-m,-k}(X; {}^0\Omega^{1/2})$ such that we have $\mathrm{id} - AB \in \Psi_0^{-\infty,0}(X; {}^0\Omega^{1/2})$ and $\mathrm{id} - BA \in \Psi_0^{-\infty,0}(X; {}^0\Omega^{1/2})$. The element B is uniquely determined up to $\Psi_0^{-\infty,-k}(X; {}^0\Omega^{1/2})$ and is called a symbolic parametrix for A. The operator A itself is called elliptic.

The proof of the following Lemma reduces easily to a symbolic argument due to Hörmander [35, Proposition 2.2.2].

PROPOSITION 2.3.8. Let $A \in \Psi_0^0(X; {}^0\Omega^{1/2})$ be with ${}^0\sigma^{(0)}(A) > 0$ and $A = A^\sharp$. Then there exists $B \in \Psi_0^0(X; {}^0\Omega^{1/2})$ with $B = B^\sharp$ and $R \in \Psi_0^{-\infty}(X; {}^0\Omega^{1/2})$ such that $A = B^2 + R$.

2.4. The 0-calculus with bounds

We are now going to enlarge the small 0-calculus by admitting kernels with non-trivial behavior at the boundary hyperfaces of $X_{0,e}^2$. As in the b-calculus, this enlarged 0-calculus is sufficient for a Fredholm theory of 0-pseudodifferential operators. This is the topic of Chapter 6.

DEFINITION 2.4.1. Let Z be a compact manifold with corners, $\mathcal{H} \subseteq \mathcal{M}_1(Z)$, $\varrho_H \in \mathcal{C}^\infty(Z)$ a defining function for $H \in \mathcal{M}_1(Z) \setminus \mathcal{H}$, and $\gamma : \mathcal{M}_1(Z) \setminus \mathcal{H} \to \mathbb{R}$ be arbitrary. Let $\mathcal{V}_\mathcal{H}(Z)$ be the space of all smooth vector fields on Z that are tangent to all $H \in \mathcal{M}_1(Z) \setminus \mathcal{H}$, and denote by

$$(2.13) \qquad \mathcal{A}_\mathcal{H}^\gamma(Z) := \{u \in \mathcal{C}^{-\infty}(Z) : (\mathcal{V}_\mathcal{H}(Z))^\ell u \subseteq \varrho_{\mathcal{M}_1 \setminus \mathcal{H}}^\gamma L^\infty(Z) \text{ for all } \ell \in \mathbb{N}_0\}$$

the space of *conormal functions* on Z with weight γ, that extend smoothly to all faces $H \in \mathcal{H}$. Here $\varrho_{\mathcal{M}_1 \setminus \mathcal{H}}^\gamma$ stands for the function $\prod_{H \in \mathcal{M}_1(Z) \setminus \mathcal{H}} \varrho_H^{\gamma(H)}$.

REMARK 2.4.2. The definition of the space $\mathcal{A}_\mathcal{H}^\gamma(Z)$ is clearly independent of the particular choice of the boundary defining functions ϱ_H. Note that by Sobolev's embedding result the elements of $\mathcal{A}_\mathcal{H}^\gamma(Z)$ are smooth in the interior of Z, i.e. $\mathcal{A}_\mathcal{H}^\gamma(Z)$ consists of all $\mathcal{C}^\infty$-functions on the interior of Z that extend smoothly to all faces $H \in \mathcal{H}$, and where the behavior at the boundary faces $H \in \mathcal{M}_1(Z) \setminus \mathcal{H}$ is controlled by the weigth system γ. If $\mathcal{H} = \emptyset$, we simply write $\mathcal{A}^\gamma(Z)$ instead of $\mathcal{A}_\emptyset^\gamma(Z)$.

Since $\mathcal{V}_\mathcal{H}(Z)$ is finitely generated as a $\mathcal{C}^\infty(Z)$-module, the right-hand side in (2.13) can be replaced by a space with only a countable number of conditions; taking the best constants leads to a Fréchet-topology on $\mathcal{A}_\mathcal{H}^\gamma(H)$ that does not depend on any choices by the open mapping theorem.

A map $\gamma : \{T_{0,e}, B_{0,e}, F_{0,e}\} = \mathcal{M}_1(X_{0,e}^2) \setminus \{\mathrm{ff}^{0,e}\} \to \mathbb{R}$ is said to be a *weight system for* X. We also use the notation $\gamma = (\gamma_T, \gamma_B, \gamma_F)$ for a weight system γ.

DEFINITION 2.4.3. Let γ be a weight system for X and $k \in \mathbb{R}$. We write $\mathcal{K}_0^{-\infty, k, \gamma}(X_{0,e}^2, KD_{0,e}^{1/2})$ for the space of all sections

$$k \in \varrho_{\mathrm{ff}^{0,e}}^{-k} \mathcal{A}_{\mathrm{ff}^{0,e}}^\gamma(X_{0,e}^2) \otimes_{\mathcal{C}^\infty(X_{0,e}^2)} \mathcal{C}^\infty(X_{0,e}^2, KD_{0,e}^{1/2}).$$

The space of all bounded linear operators $A : \dot{\mathcal{C}}^\infty(X, {}^0\Omega^{1/2}) \to \mathcal{C}^{-\infty}(X, {}^0\Omega^{1/2})$ with lifted Schwartz kernel κ_A in $\mathcal{K}_0^{-\infty, k, \gamma}(X_{0,e}^2, KD_{0,e}^{1/2})$ is denoted by $\Psi_0^{-\infty, k, \gamma}(X; {}^0\Omega^{1/2})$. Finally,

$$\Psi_0^{m, k, \gamma}(X; {}^0\Omega^{1/2}) := \Psi_0^{m, k}(X; {}^0\Omega^{1/2}) + \Psi_0^{-\infty, k, \gamma}(X; {}^0\Omega^{1/2})$$

is the space of mth-*order 0-pseudodifferential operators of type k with bounds* γ.

For the convenience of the reader, and since we need it later on anyway, let us give a description of the lifted Schwartz kernel of $A \in \Psi_0^{-\infty, 0, \gamma}(X; {}^0\Omega^{1/2})$ in local coordinates (τ, U, r, y') as in (1.8) near the extended 0-front face $\mathrm{ff}^{0,e}$. Let $\kappa_A = \widehat{\kappa}_A(\tau, U, r, y') \left| d\tau \, dU \, \frac{dr}{r^n} dy' \right|^{\frac{1}{2}}$ be a local representation of the lifted Schwartz kernel of A. Then we have $\widehat{\kappa}_A \in \mathcal{C}^\infty((-1,1)) \widehat{\otimes}_\pi \mathcal{C}^\infty(\mathbb{R}_U^{n-1}) \widehat{\otimes}_\pi \mathcal{C}^\infty(\overline{\mathbb{R}}_+) \widehat{\otimes}_\pi \mathcal{C}_c^\infty(\mathbb{R}_{y'}^{n-1})$ with
$$(2.14) \qquad \left| [(1 - \tau^2) \partial_\tau]^j \partial_U^\alpha \partial_r^\ell \partial_{y'}^\beta \widehat{\kappa}_A(\tau, U, r, y') \right| \leq \mathrm{const} \, (1 + \tau)^{\gamma_T} (1 - \tau)^{\gamma_B} <U>^{-\gamma_F - |\alpha|}$$

for all $j, \ell \in \mathbb{N}_0$ and all multi-indices $\alpha, \beta \in \mathbb{N}_0^{n-1}$. For the U-part of the estimate (2.14) note that under radial compactification the space $\mathcal{A}^m(S_+^{n-1})$ corresponds to the usual symbol space $S^m(;\mathbb{R}^{n-1})$.

2.5. Basic properties of the 0-calculus with bounds

PROPOSITION 2.5.1. *Let $\gamma = (\gamma_T, \gamma_B, \gamma_F)$ be an arbitrary weight system. Then we have for all $z \in \mathbb{C}$*
$$\varrho_N^z \Psi_0^{-\infty,k,\gamma}(X; {}^0\Omega^{1/2}) \varrho_N^{-z} = \Psi_0^{-\infty,k,\gamma_z}(X; {}^0\Omega^{1/2})$$
where $\gamma_z = ((\gamma_z)_T, (\gamma_z)_B, (\gamma_z)_F) = (\gamma_T + \operatorname{Re} z, \gamma_B - \operatorname{Re} z, \gamma_F)$.

Proof: It suffices to note that the lifted Schwartz kernel κ_B of $B := \varrho_N^z A \varrho_N^{-z}$ equals $\varrho_{T_{0,e}}^z \varrho_{B_{0,e}}^{-z} \kappa_A$ up to smooth non-vanishing factors. □

Note that the conormal behavior at the face $F_{0,e}$ is not affected by conjugation with ϱ_N^z.

2.6. The indicial function

As for the b-calculus, the homogeneous principal symbol does not suffice to characterize the Fredholm property of 0-pseudodifferential operators. The additional, infinite-dimensional and non-commutative symbol, the *reduced normal operator* is considered in Section 4; however, to understand the reduced normal operator completely, we need the *indicial function* of a 0-pseudodifferential operator. It is defined similar to the indicial family of a b-pseudodifferential operator, and is, in fact, the indicial family of the b-part of the reduced normal operator. For the definition of the indicial function we follow [65]. For simplicity, we start with operators $A \in \Psi_0^m(X; {}^0\Omega^{1/2})$ in the small 0-calculus.

Fix a boundary defining function $\varrho_N : X \to \overline{\mathbb{R}}_+$ for X. Then ϱ_N yields a restriction map
$$\mathcal{R}_{\varrho_N} : \mathcal{C}^\infty(X, {}^0\Omega^{1/2}) \longrightarrow \mathcal{C}^\infty(\partial X, \Omega^{1/2}).$$
Because of Lemma 2.3.3, we have $\varrho_N^{-1} A \varrho_N^1 \in \Psi_0^m(X; {}^0\Omega^{1/2})$, thus, the action of A on $\mathcal{C}^\infty(\partial X, \Omega^{1/2})$ given by
$$A_\partial f = \mathcal{R}_{\varrho_N}(AF) \in \mathcal{C}^\infty(\partial X, \Omega^{1/2}) \text{ if } F \in \mathcal{C}^\infty(X, {}^0\Omega^{1/2}) \text{ satisfies } \mathcal{R}_{\varrho_N}(F) = f$$
is well-defined.

LEMMA 2.6.1. *For any $A \in \Psi_0^m(X; {}^0\Omega^{1/2})$, there exists a $a \in \mathcal{C}^\infty(\partial X)$ such that $A_\partial f = af$ for all $f \in \mathcal{C}^\infty(\partial X, \Omega^{1/2})$. The map*
$$(\cdot)_\partial : \Psi_0^\infty(X; {}^0\Omega^{1/2}) \longrightarrow \mathcal{C}^\infty(\partial X) : A \longmapsto a$$
is continuous, satisfies $(AB)_\partial = A_\partial B_\partial$ for $A, B \in \Psi_0^\infty(X; {}^0\Omega^{1/2})$, and is called restriction to the boundary.

Proof: First note that for $q \in \partial X$ we have $A_\partial f(q) = 0$, whenever $f(q) = 0$. Indeed, choose local coordinates (x, y) near q with $(x, y)(q) = (0, 0)$. A Taylor expansion near q yields $f = \sum_{j=1}^{n-1} y_j f_j$, thus, we obtain for any F with $F(x, y) = f(y)$ for small x
$$AF = \sum_{j=1}^{n-1} y_j A f_j + \sum_{j=1}^{n-1} [A, y_j] f_j.$$
Since the lifted Schwartz kernel κ_j of $[A, y_j]$ is given in the local coordinates (1.8) by $\kappa_j = -rU_j\kappa_A$, i.e. a kernel vanishing at the extended 0-front face $\mathrm{ff}^{0,e}$, we have $(A_\partial f)(q) = AF(q) = 0$ as desired.

Another Taylor expansion shows that this can only happen when A_∂ is given by multiplication with a smooth function. This completes the proof. □

DEFINITION 2.6.2. For $A \in \Psi_0^m(X; {}^0\Omega^{1/2})$, we call
$$I_A : \mathbb{C} \longrightarrow \mathcal{C}^\infty(X) : z \longmapsto (\varrho_N^{-iz} A \varrho_N^{iz})_\partial$$
the *indicial function* of A.

The main properties of the indicial function are summarized in the following Proposition.

PROPOSITION 2.6.3. *Let $A, B \in \Psi_0^\infty(X; {}^0\Omega^{1/2})$ be arbitrary. Then we have:*
(a) $I_A \in \mathcal{O}(\mathbb{C}, \mathcal{C}^\infty(\partial X))$.
(b) $I_{AB} = I_A I_B$.
(c) $I_{A^\sharp}(z) = \overline{I_A(\bar{z} - i(1 - n))}$ *for* $z \in \mathbb{C}$.
(d) *With respect to local coordinates (τ, U, r, y') as in (1.8) the indicial function I_A is given by*

$$(2.15) \quad I_A(z)(y) = \sqrt{2} \int_{\mathbb{R}_U^{n-1}} \int_{-1}^{1} \left(\frac{1+\tau}{1-\tau}\right)^{-i(z+i\frac{n-1}{2})} \widehat{\kappa}_A(\tau, U, 0, y) \frac{d\tau}{(1-\tau^2)^{1/2}} dU$$

for $y \in \partial X$ and $z \in \mathbb{C}$. Here, $\kappa_A = \widehat{\kappa}_A(\tau, U, r, y') |d\tau \, dU \frac{dr}{r^n} dy'|^{\frac{1}{2}}$ is a representation of the lifted Schwartz kernel κ_A of A in local coordinates near the front face.
(e) *For $A \in \Psi_0^m(X; {}^0\Omega^{1/2})$, we have*
$$[\partial X \times \mathbb{R}_\lambda \ni (q, \lambda) \longmapsto I_A(\lambda + i\mu)(q)] \in \mathcal{C}^\infty(\partial X, S_{cl}^m(; \mathbb{R}_\lambda))$$
with uniform estimates for μ in compact subsets of $\mathbb{R}$.
(f) *The indicial function $I_A(z)$ does not depend on the choice of the defining function ϱ_N.*

Proof: (a) and (b) follow immediately from Lemma 2.6.1 and Definition 2.6.2, (c) can be checked using for instance (2.15), (d) is a straightforward consequence of the formula for the action of 0-pseudodifferential operators in local coordinates, and (e) follows from a combination of (2.15) with standard techniques from oscillatory integrals. By Lemma 2.6.1, $(\cdot)_\partial$ commutes with the multiplication by functions; this implies (f). □

2.7. General bundles

The 0-calculus extends naturally to arbitrary vector bundles over X. Indeed, let $E, F \to X$ be two finite-dimensional complex vector bundles. Then we define $\Psi_0^{m,k,\gamma}(X; E, F)$ to be the space
$$(2.16)$$
$$\Psi_0^{m,k,\gamma}(X; {}^0\Omega^{1/2}) \otimes_{\mathcal{C}^\infty(X_{0,e}^2)} \mathcal{C}^\infty(X_{0,e}^2, (\beta_{0,e}^2)^*(\text{Hom}(E \otimes {}^0\Omega^{-1/2}, F \otimes {}^0\Omega^{-1/2}))).$$

As one might expect, most of the properties of 0-pseudodifferential operators remain true with the obvious modifications. For instance, a 0-pseudodifferential operator $A \in \Psi_0^{m,k}(X; E, F)$ induces continuous linear operators $A : \dot{\mathcal{C}}^\infty(X, E) \to \dot{\mathcal{C}}^\infty(X, F)$ and $A : \mathcal{C}^{-\infty}(X, E) \to \mathcal{C}^{-\infty}(X, F)$.

CHAPTER 3

The b-c-calculus on an interval

Let M be a compact manifold whose boundary has more than one component. Then we can consider Lie algebras of smooth vector fields on M that behave differently at different components of the boundary, and construct a corresponding pseudodifferential calculus. For understanding the reduced normal operator of 0-pseudodifferential operators we need in fact a calculus on a closed interval with b-behavior near one end-point, and a cusp-behavior near the other. Since the b-calculus as well as the cusp-calculus are well-known, we simply give a brief outline of the construction of the b-c-calculus on $M = [0, 1]$, and summarize the properties that are essential for us. For more details on the b- resp. c–calculus, we refer to [**32, 50, 55, 64, 68**] resp. [**45, 60, 72**].

Throughout this chapter M stands for the compact interval $[0, 1]$. Let us denote by ϱ_0 resp. ϱ_1 defining functions for the components $\{0\}$ resp. $\{1\}$ of ∂M. Of course, we can take $\varrho_0 : x \mapsto x$ and $\varrho_1 : x \mapsto 1 - x$.

3.1. The b-c-structure algebra

DEFINITION 3.1.1. A smooth vector field V on M is called a *b-c-vector field* on M provided $Vf \in \varrho_0 \varrho_1^2 \mathcal{C}^\infty(M)$ for all $f \in \mathcal{C}^\infty(M)$. We write $\mathcal{V}_{b,c}(M)$ for the Lie-algebra of all b-c-vector fields on M.

Note that the definition does not depend on the choice of the defining functions ϱ_0 and ϱ_1. A smooth vector field on M is a b-c-vector field, if and only if it is of the form $V = a(x)x(1-x)^2 \partial_x$ with $a \in \mathcal{C}^\infty(M)$. Thus, there exists a rank 1 smooth vector bundle $^{b,c}TM$ on M together with a natural map $j^{b,c} : {}^{b,c}TM \to TM$ such that $\mathcal{V}_{b,c}(M) = j^{b,c}(\mathcal{C}^\infty(M, {}^{b,c}TM))$; it is called the *$b$-$c$-tangent bundle*, its dual, $^{b,c}T^*M$, the *b-c-cotangent bundle*.

3.2. The b-c-double space

As for the 0-calculus, we have to replace the space M^2 by a blown-up version $M^2_{b,c}$ for a convenient definition of the b-c-pseudodifferential calculus. Again, we start with the b-blow-up

$$\beta_b^2 : M_b^2 := [M^2; \{(0,0)\} \cup \{(1,1)\}] \longrightarrow M^2.$$

Let $\text{ff}^b(j)$, $j = 0, 1$, be the new boundary faces obtained by blowing up the points (j,j), $\Delta_b := \overline{(\beta_b^2)^{-1}(\Delta \setminus \partial M^2)}^{M_b^2}$ the b-diagonal, and $B := \Delta_b \cap \text{ff}^b(1)$. Then the composition of the maps

$$\beta_{b,c}^2 : M_{b,c}^2 := [M_b^2; B] \longrightarrow M_b^2 \xrightarrow{\beta_b^2} M^2$$

is called the *b-c-blow up*, and $M_{b,c}^2$ the *b-c-double space*; it is a compact manifold with corners up to codimension 2. From the eight boundary hyperfaces of $M_{b,c}^2$, only four are of particular interest for us: the *b-front face* $\mathrm{ff}^b(0)$, the lift $\mathrm{lb} = \mathrm{lb}(0)$ resp. $\mathrm{rb} = \mathrm{rb}(0)$ of the *left* resp. *right boundary faces* of M_b^2 corresponding to $(0,0)$, and the *cusp front face* ff^c, i.e. the new boundary face produced by the last blow-up. Moreover, we need the *b-c-diagonal* $\Delta_{b,c} := \overline{(\beta_{b,c}^2)^{-1}(\Delta \setminus \partial M^2)}^{M_{b,c}^2}$.

To give another interpretation of the b-c-double-space, let

$$\mathrm{RC} : \overline{\mathbb{R}}_+ \longrightarrow S_{++}^1 := \{\omega_1, \omega_2\} \in S^1 : \omega_1 \geq 0, \omega_2 \geq 0\} : x \longmapsto \frac{(1,x)}{|(1,x)|}$$

be the *radial compactification of the half-axis* $\overline{\mathbb{R}}_+$. It is often convenient to identify S_{++}^1 with $M = [0,1]$ via $(\omega_1, \omega_2) \mapsto 2/\pi \arctan(\omega_2/\omega_1)$. The radial compactification then becomes

$$(3.1) \qquad \mathrm{RC} : \overline{\mathbb{R}}_+ \longrightarrow M : x \mapsto 2/\pi \arctan x =: z.$$

Note that the point at infinity corresponds to $\omega_1 = 0$ resp. $z = 1$. Recall the basic *b*-blow-down map

$$(3.2) \qquad \beta_b^2 : [-1,1] \times \overline{\mathbb{R}}_+ \longrightarrow \overline{\mathbb{R}}_+ \times \overline{\mathbb{R}}_+ : (\tau, r) \longmapsto \left(\frac{r}{2}(1+\tau), \frac{r}{2}(1-\tau)\right),$$

and let us define the blow ups

$$\beta : M_N := \big[\, [-1,1] \times [0,1]; \{(0,1)\} \,\big] \longrightarrow [-1,1] \times [0,1],$$
$$\beta : \widetilde{M}_b^2 := \big[\, [-1,1] \times [0,1]; \{(-1,1)\} \cup \{(+1,1)\} \,\big] \longrightarrow [-1,1] \times [0,1], \text{ and}$$
$$\beta : \widetilde{M}_{b,c}^2 := \big[\, \widetilde{M}_b^2; \{(0,1)\} \,\big] \longrightarrow \widetilde{M}_b^2.$$

For the sake of completeness, let us denote by ff^{M_N} the boundary face of M_N that is obtained by blowing up $\{(0,1)\}$, by lb resp. rb the lift of $\{-1\} \times [0,1]$ resp. $\{+1\} \times [0,1]$, by $F_{M_N,+}$ resp. $F_{M_N,-}$ that of $[-1,0] \times \{+1\}$ resp. $[0,1] \times \{+1\}$, and by $\Delta_{M_N} := \overline{\beta^{-1}(\{0\} \times [0,1))}^{M_N}$ that of $\{\tau = 0\}$. We are mainly interested in the relation between the spaces $M_{b,c}^2$ and M_N. Consider now the following diagram

$$(3.3) \qquad \begin{array}{c}
\overline{\mathbb{R}}_+ \times \overline{\mathbb{R}}_+ \xleftarrow{\beta_b^2} [-1,1] \times \overline{\mathbb{R}}_+ \xrightarrow{\mathrm{id} \times \mathrm{RC}} [-1,1] \times [0,1] \\
\mathrm{RC} \times \mathrm{RC} \downarrow \qquad\qquad\qquad\qquad\qquad \uparrow \beta \quad\quad \nwarrow \beta \\
M^2 \xleftarrow{\beta_b^2} M_b^2 \dashrightarrow^{\varphi_b} \widetilde{M}_b^2 \quad\quad \# \quad\quad M_N \\
\uparrow \beta \qquad\qquad \uparrow \beta \qquad\qquad \nearrow \beta \\
M_{b,c}^2 \dashrightarrow^{\varphi_{b,c}} \widetilde{M}_{b,c}^2
\end{array}$$

For the existence of the mappings φ_b and $\varphi_{b,c}$, note that all other arrows in (3.3) are diffeomorphisms when restricted to the interior part $\mathbb{R}_+ \times \mathbb{R}_+$ that is basically not affected by the blow down maps and the radial compactification. The commutativity of the right triangle up to canonical diffeomorphisms follows from the commutativity of the blow up because the pieces to be blown up are disjoint.

PROPOSITION 3.2.1. *The maps φ_b and $\varphi_{b,c}$ extend from the interior to diffeomorphisms $\varphi_b : M_b^2 \to \widetilde{M}_b^2$ and $\varphi_{b,c} : M_{b,c}^2 \to \widetilde{M}_{b,c}^2$.*

Proof: Since going from M_b^2 to $M_{b,c}^2$ corresponds to the same blow up as in the step from $\widetilde{M}_b^2$ to $\widetilde{M}_{b,c}^2$, it suffices to consider φ_b. In that case, a straightforward computation in local coordinates shows that φ extends from the interior to a diffeomorphism. $\square$

3.3. b-c-densities

By applying the functor Ω^α of α-densities to the b-c-tangent bundle $^{b,c}TM$, we obtain the bundle $^{b,c}\Omega^\alpha(M)$ of *b-c-α-densities*. A non vanishing section of $^{b,c}\Omega^\alpha(M)$ is given by $\left|\frac{dz}{z(1-z)^2}\right|^\alpha$. In particular, we obtain a natural scalar product on $\dot{\mathcal{C}}^\infty(M, {}^{b,c}\Omega^{1/2})$ by

$$< f_1, f_2 >_{L^2(M, {}^{b,c}\Omega^{1/2})} := \int_M f_1 \bar{f}_2 = \int_0^1 \widehat{f}_1(z)\overline{\widehat{f}_2(z)} \frac{dz}{z(1-z)^2}$$

for $f_j = \widehat{f}_j \left|\frac{dz}{z(1-z)^2}\right|^{\frac{1}{2}} \in \dot{\mathcal{C}}^\infty(M, {}^{b,c}\Omega^{1/2})$. Let $L^2(M, {}^{b,c}\Omega^{1/2})$ be the Hilbert space completion of $\dot{\mathcal{C}}^\infty(M, {}^{b,c}\Omega^{1/2})$ with respect to this scalar product. On the b-c-double space $M_{b,c}^2$, we use the *b-c-kernel half density bundle* $KD_{b,c}^{1/2} := \varrho_{\mathrm{ff}^b}^{-1/2} \varrho_{\mathrm{ff}^c}^{-1} \Omega^{1/2}(M_{b,c}^2)$. Then the blow-down map $\beta_{b,c}^2 : M_{b,c}^2 \to M^2$ induces via pull-back and duality isomorphisms

$$(\beta_{b,c}^2)^* : \dot{\mathcal{C}}^\infty(M^2, {}^{b,c}\Omega^{1/2} \boxtimes {}^{b,c}\Omega^{1/2}) \xrightarrow{\cong} \dot{\mathcal{C}}^\infty(M_{b,c}^2, KD_{b,c}^{1/2}) \text{, and}$$

$$(3.4) \quad (\beta_{b,c}^2)_* : \quad \mathcal{C}^{-\infty}(M_{b,c}^2, KD_{b,c}^{1/2}) \xrightarrow{\cong} \mathcal{C}^{-\infty}(M^2, {}^{b,c}\Omega^{1/2} \boxtimes {}^{b,c}\Omega^{1/2}),$$

and the Schwartz kernel theorem reads as follows: There is a one-to-one correspondence between the continuous linear operators $\dot{\mathcal{C}}^\infty(M, {}^{b,c}\Omega^{1/2}) \to \mathcal{C}^{-\infty}(M, {}^{b,c}\Omega^{1/2})$ on the one hand, and the space of *extendible distributions* $\mathcal{C}^{-\infty}(M_{b,c}^2, KD_{b,c}^{1/2})$ on the other hand. Again, we call the distribution $\kappa_A \in \mathcal{C}^{-\infty}(M_{b,c}^2, KD_{b,c}^{1/2})$ corresponding to $A : \dot{\mathcal{C}}^\infty(M, {}^{b,c}\Omega^{1/2}) \to \mathcal{C}^{-\infty}(M, {}^{b,c}\Omega^{1/2})$ under this correspondence the *lifted Schwartz kernel* of A.

3.4. The b-c-calculus with bounds

The b-c-calculus with bounds consists of three parts; let us start with the small calculus. As in Section 2, we define the *small b-c-calculus* $\Psi_{b,c}^m(M; {}^{b,c}\Omega^{1/2})$ of order $m \in \mathbb{C}$ using the Schwartz kernel theorem by

$$\Psi_{b,c}^m(M; {}^{b,c}\Omega^{1/2}) \longleftrightarrow \underbrace{\left\{\kappa \in I_{cl}^m(M_{b,c}^2, \Delta_{b,c}; KD_{b,c}^{1/2}) : \kappa \equiv 0 \text{ at } \partial M_{b,c}^2 \setminus (\mathrm{ff}^b \cup \mathrm{ff}^c)\right\}}_{\subseteq \mathcal{C}^{-\infty}(M_{b,c}^2, KD_{b,c}^{1/2})}.$$

Observe that each b-c-operator $A \in \Psi_{b,c}^m(M; {}^{b,c}\Omega^{1/2})$ induces continuous linear maps $A : \dot{\mathcal{C}}^\infty(M, {}^{b,c}\Omega^{1/2}) \to \dot{\mathcal{C}}^\infty(M, {}^{b,c}\Omega^{1/2})$ and $A : \mathcal{C}^\infty(M, {}^{b,c}\Omega^{1/2}) \to \mathcal{C}^\infty(M, {}^{b,c}\Omega^{1/2})$. Since we get the lifted kernel of $\varrho_0^{z_0} \varrho_1^{z_1} A \varrho_1^{-z_1} \varrho_0^{-z_0}$, $z_0, z_1 \in \mathbb{C}$, by multiplying the lifted kernel of A with a function that is 1 at $\Delta_{b,c}$, smooth up to $\mathrm{ff}^b(0)$ and ff^c, and polynomially bounded at all other boundary faces of $M_{b,c}^2$, we have

$$(3.5) \quad \varrho_0^{z_0} \varrho_1^{z_1} \Psi_{b,c}^m(M; {}^{b,c}\Omega^{1/2}) \varrho_1^{-z_1} \varrho_0^{-z_0} = \Psi_{b,c}^m(M; {}^{b,c}\Omega^{1/2}).$$

This allows to define b-c-operators with a more general behavior at the faces $\mathrm{ff}^b(0)$ and ff^c. For notational simplicity, and since this is the only extension that we really need, we restrict ourselves to the cusp front face ff^c.

DEFINITION 3.4.1. For $k \in \mathbb{C}$, we call
$$\Psi_{b,c}^{m,k}(M; {}^{b,c}\Omega^{1/2}) := \varrho_1^{-k} \Psi_{b,c}^m(M; {}^{b,c}\Omega^{1/2}) = \Psi_{b,c}^m(M; {}^{b,c}\Omega^{1/2}) \varrho_1^{-k}$$
the *extended b-c-calculus*.

We are going to add operators of symbolic order $-\infty$ but with non-trivial behavior at the left (lb) and right (rb) boundary of the double space $M_{b,c}^2$. Let $\mathcal{H} := \mathcal{M}_1(M_{b,c}^2) \setminus \{\mathrm{lb}, \mathrm{rb}\}$, and $\gamma = (\gamma_{\mathrm{lb}}, \gamma_{\mathrm{rb}}) : \mathcal{M}_1(M_{b,c}^2) \setminus \mathcal{H} \to \mathbb{R}$ be a *weight system* for M. Let $\mathcal{K}^{-\infty,\gamma}(M_{b,c}^2; KD_{b,c}^{1/2})$ be the space of all sections
$$\kappa \in \mathcal{A}_{\mathcal{H}}^{\gamma}(M_{b,c}^2) \otimes_{\mathcal{C}^\infty(M_{b,c}^2)} \mathcal{C}^\infty(M_{b,c}^2, KD_{b,c}^{1/2})$$
that vanish with all derivatives at all hyperfaces $H \in \mathcal{M}_1(M_{b,c}^2) \setminus (\mathrm{ff}^b, \mathrm{lb}, \mathrm{rb})$, and let us denote by $\widetilde{\Psi}_{b,c}^{-\infty,-\infty,\gamma}(M, {}^{b,c}\Omega^{1/2})$ the space of all continuous linear maps $A : \dot{\mathcal{C}}^\infty(M, {}^{b,c}\Omega^{1/2}) \to \mathcal{C}^{-\infty}(M, {}^{b,c}\Omega^{1/2})$ whose lifted Schwartz kernel κ_A belongs to $\mathcal{K}^{-\infty,\gamma}(M_{b,c}^2; KD_{b,c}^{1/2})$.

To define the last contribution to the b-c-calculus with bounds, first note that M^2 has four boundary hyperfaces that we label with $\mathrm{lb}(j) = \{z = j\}$ and $\mathrm{rb}(j) = \{z' = j\}$, $j = 0, 1$. Let $\mathcal{H} = \{\mathrm{lb}(1), \mathrm{rb}(1)\}$, $\gamma : \mathcal{M}_1(M^2) \setminus \mathcal{H} \to \mathbb{R}$, and $\Psi^{-\infty,\gamma}(M, {}^{b,c}\Omega^{1/2})$ the space of operators $A : \dot{\mathcal{C}}^\infty(M, {}^{b,c}\Omega^{1/2}) \to \mathcal{C}^{-\infty}(M, {}^{b,c}\Omega^{1/2})$ whose Schwartz kernels are sections $k \in \mathcal{A}_{\mathcal{H}}^{\gamma}(M^2) \otimes_{\mathcal{C}^\infty(M^2)} \mathcal{C}^\infty(M^2, {}^{b,c}\Omega^{1/2} \boxtimes {}^{b,c}\Omega^{1/2})$ that vanish to infinite order at the faces $\mathrm{lb}(1)$ and $\mathrm{rb}(1)$.

DEFINITION 3.4.2. For $m \in \mathbb{C}$, $k \in \mathbb{C}$, and $\gamma = (\gamma_{\mathrm{lb}}, \gamma_{\mathrm{rb}})$ we call the sum
$$\Psi_{b,c}^{m,k,\gamma}(M; {}^{b,c}\Omega^{1/2}) := \Psi_{b,c}^{m,k}(M; {}^{b,c}\Omega^{1/2}) + \widetilde{\Psi}_{b,c}^{-\infty,-\infty,\gamma}(M, {}^{b,c}\Omega^{1/2}) +$$
$$+ \Psi^{-\infty,\gamma'}(M, {}^{b,c}\Omega^{1/2})$$
the *extended b-c-calculus of order m with bounds γ*. Here, γ' stands for the weight $\gamma' := (\gamma'_{\mathrm{lb}}, \gamma'_{\mathrm{rb}}) = (\gamma_{\mathrm{lb}} + 1/2, \gamma_{\mathrm{rb}} + 1/2)$.

Note that $\Psi_{b,c}^{m,k}(M; {}^{b,c}\Omega^{1/2})$ formally corresponds to $\Psi_{b,c}^{m,k,\gamma}(M; {}^{b,c}\Omega^{1/2})$ with weights $\gamma_{\mathrm{lb}} := \gamma_{\mathrm{rb}} := \infty$. The space $\Psi_{b,c}^{m,k,\gamma}(M; {}^{b,c}\Omega^{1/2})$ carries naturally the Fréchet topology of a non-direct sum of Fréchet spaces.

3.5. Basic properties of the b-c-calculus

Let us start with the homogeneous principal symbol map; it is induced by the homogeneous principal symbol map for classical conormal distributions, and leads to a map $\Psi_{b,c}^{m,k,\gamma}(M; {}^{b,c}\Omega^{1/2}) \to \varrho_1^{-k} S^{[m]}({}^{b,c}T^*M)$. Fixing ϱ_1 and a defining function ϱ_σ for the b-c-cosphere bundle ${}^{b,c}S^*M := ({}^{b,c}T^*M \setminus \{0\})/\mathbb{R}_+$ in the radial compactification ${}^{b,c}\overline{T}^*M$ of ${}^{b,c}T^*M$ yields an identification
$$\varrho_1^{-k} S^{[m]}({}^{b,c}T^*M) \longleftrightarrow \mathcal{C}^\infty({}^{b,c}S^*M).$$
Note, however, that ${}^{b,c}S^*M = M \times \{\pm 1\}$ because the fibers of ${}^{b,c}T^*M$ are one-dimensional. The corresponding induced morphism
$${}^{b,c}\sigma^{(m,k)} = ({}^{b,c}\sigma^{(m,k)}(-1), {}^{b,c}\sigma^{(m,k)}(+1)) : \Psi_{b,c}^{m,k,\gamma}(M; {}^{b,c}\Omega^{1/2}) \to \mathcal{C}^\infty(M) \oplus \mathcal{C}^\infty(M)$$

3.5. BASIC PROPERTIES OF THE b-c-CALCULUS

is called the *homogeneous principal b-c-symbol*. To understand Fredholm properties of b-c-operators, we need two more invariants, the b- resp. c-*indicial families*. For $\gamma = (\gamma_{\mathrm{lb}}, \gamma_{\mathrm{rb}})$ with $\gamma_{\mathrm{lb}} > -1/2$ and $\gamma_{\mathrm{rb}} > -1/2$, let $(\cdot)_{\partial_j} : \Psi_{b,c}^{m,k,\gamma}(M; {}^{b,c}\Omega^{1/2}) \to \mathbb{C}$, $j=0,1$, be the morphisms of restricting to the boundary component $\{0\}$ resp. $\{1\}$ of M given by $A_{\partial_j} = \mathcal{R}_{\varrho_j}(AF)$ for any $F \in \mathcal{C}^\infty(M, {}^{b,c}\Omega^{1/2})$ with $\mathcal{R}_{\varrho_j}(F) = 1$; here, the restriction $\mathcal{R}_{\varrho_j} : \mathcal{C}^\infty(M, {}^{b,c}\Omega^{1/2}) \to \mathbb{C}$ is induced by the defining function ϱ_j for $\{j\}$. For $j=1$, we have, of course, to assume additionally $k=0$. The morphism $(\cdot)_{\partial_j}$ does not depend on the choice of the defining function ϱ_j. For $A \in \Psi_{b,c}^{m,k,\gamma}(M; {}^{b,c}\Omega^{1/2})$, we call

$$I_b(A)(z) := (\varrho_0^{-iz} A \varrho_0^{iz})_{\partial_0} \text{ for } z \in \mathbb{C} \text{ with } -\frac{1}{2} - \gamma_{\mathrm{rb}} < \operatorname{Im} z < \frac{1}{2} + \gamma_{\mathrm{lb}} \text{ resp.}$$

$$I_c^{(k)}(A)(\xi) := (e^{i\xi/\varrho_1} \varrho_1^k A e^{-i\xi/\varrho_1})_{\partial_1} = (e^{i\xi/\varrho_1} A \varrho_1^k e^{-i\xi/\varrho_1})_{\partial_1} \text{ for } \xi \in \mathbb{R}_\xi$$

the b- resp. c-*indicial family* of A. The b-indicial family obviously does not depend on the parameter k.

REMARK 3.5.1. Observe that $I_c^{(k)}(A) \in S_{cl}^m(;\mathbb{R}_\xi)$ and that $I_b(A)$ belongs to the space of all analytic functions $h : \{-1/2 - \gamma_{\mathrm{rb}} < \operatorname{Im} z < 1/2 + \gamma_{\mathrm{lb}}\} \to \mathbb{C}$ such that $[\xi \mapsto h(\xi + i\mu)] \in S_{cl}^m(;\mathbb{R}_\xi)$ with uniform estimates for μ in compact subsets of $(-1/2 - \gamma_{\mathrm{rb}}, 1/2 + \gamma_{\mathrm{lb}})$. In particular, for $A \in \Psi_{b,c}^m(M; {}^{b,c}\Omega^{1/2})$, we have $I_b(A) \in M_{\mathcal{O}}^m$, the space of all entire maps h such that $[\xi \mapsto h(\xi + i\mu)] \in S_{cl}^m(;\mathbb{R}_\xi)$ with uniform estimates for μ in compact subsets of $\mathbb{R}$.

For the convenience of the reader, let us give a local description of the b- resp. c-indicial family of an operator $A \in \Psi_{b,c}^{m,k,\gamma}(M; {}^{b,c}\Omega^{1/2})$. If $\widehat{\kappa}_A(\tau, r)\left|\frac{dr}{r} d\tau\right|^{\frac{1}{2}}$ resp. $\widehat{\kappa}_{A_1}(S, x') {x'}^{-k} \left|dS \frac{dx'}{x'}\right|^{\frac{1}{2}}$ is the a local representation of the lifted Schwartz kernel of $A = A_1 \varrho_1^{-k} \in \Psi_{b,c}^{m,k,\gamma}(M; {}^{b,c}\Omega^{1/2})$ with respect to local coordinates

$$\tau = \frac{x - x'}{x + x'} \text{ and } r = x + x' \text{ resp. } S = \frac{1}{x'} - \frac{1}{x} \text{ and } x'$$

near the b- resp. the c-front face, then for $z \in \mathbb{C}$ with $-\frac{1}{2} - \gamma_{\mathrm{rb}} < \operatorname{Im} z < \frac{1}{2} + \gamma_{\mathrm{lb}}$ we have

$$(3.6) \qquad I_b(A)(z) = \int_{-1}^{1} \left(\frac{1+\tau}{1-\tau}\right)^{-iz} \widehat{\kappa}_A(\tau, 0) \frac{d\tau}{(1-\tau^2)^{1/2}}$$

$$(3.7) \qquad I_c^{(k)}(A)(\xi) = \int_{-\infty}^{\infty} e^{-iS\xi} \widehat{\kappa}_{A_1}(S, 0)\, dS \text{ for } \xi \in \mathbb{R}_\xi.$$

We summarize the basic symbolic properties of b-c-operators in the small calculus in the following theorem. Recall that the formal adjoint $A^\sharp$ of an operator $A : \dot{\mathcal{C}}^\infty(M, {}^{b,c}\Omega^{1/2}) \to \dot{\mathcal{C}}^\infty(M, {}^{b,c}\Omega^{1/2})$ is defined, if it exists, by the property

$$<Af, g>_{L^2(M, {}^{b,c}\Omega^{1/2})} = <f, A^\sharp g>_{L^2(M, {}^{b,c}\Omega^{1/2})} \text{ for all } f, g \in \dot{\mathcal{C}}^\infty(M, {}^{b,c}\Omega^{1/2}).$$

THEOREM 3.5.2. *For $A_j \in \Psi_{b,c}^{m_j, k_j}(M; {}^{b,c}\Omega^{1/2})$ and $A \in \Psi_{b,c}^{m,k}(M; {}^{b,c}\Omega^{1/2})$ we have $A_1 A_2 \in \Psi_{b,c}^{m_1+m_2, k_1+k_2}(M; {}^{b,c}\Omega^{1/2})$ and $A^\sharp \in \Psi_{b,c}^{m,k}(M; {}^{b,c}\Omega^{1/2})$ with*

(a) ${}^{b,c}\sigma^{(m_1+m_2, k_1+k_2)}(A_1 A_2) = {}^{b,c}\sigma^{(m_1, k_1)}(A_1) {}^{b,c}\sigma^{(m_2, k_2)}(A_2) \in \mathcal{C}^\infty(M)^2$ *and*
${}^{b,c}\sigma^{(m,k)}(A^\sharp) = \overline{{}^{b,c}\sigma^{(m,k)}(A)},$

(b) $I_b(A_1 A_2)(z) = I_b(A_1)(z) I_b(A_2)(z)$ and $I_b(A^\sharp)(z) = \overline{I_b(A)(\bar{z})}$,

(c) $I_c^{(k_1+k_2)}(A_1 A_2)(\xi) = I_c^{(k_1)}(A_1)(\xi) I_c^{(k_2)}(A_2)(\xi)$ and $I_c^{(k)}(A^\sharp) = \overline{I_c^{(k)}(A)}$,

(d) $\sigma^{(m)}(I_b(A)|_{\{\mathrm{Im}\, z = \mu\}}) = {}^{b,c}\sigma^{(m,k)}(A)|_{\partial_0}$, $\mu \in \mathbb{R}$, and

(e) $\sigma^{(m)}(I_c^{(k)}(A)) = {}^{b,c}\sigma^{(m,k)}(A)|_{\partial_1}$,

where $\sigma^{(m)} : S_{cl}^m(;\mathbb{R}) \to S^{[m]}(\mathbb{R}) \cong \mathcal{C}^\infty(\{\pm 1\}) = \mathbb{C} \oplus \mathbb{C}$ is the homogeneous principal part of a classical symbol.

Similar results with appropriate restrictions on z and γ hold also for the b-c-calculus with bounds. Let us only mention the following special case that is of importance for the proof of the Fredholm property for 0-pseudodifferential operators.

PROPOSITION 3.5.3. *Let $\gamma = (\gamma_{\mathrm{lb}}, \gamma_{\mathrm{rb}})$ be an arbitrary weight system with $\gamma_{\mathrm{lb}}, \gamma_{\mathrm{rb}} > -1/2$. Then for $A \in \Psi_{b,c}^{-\infty,-\infty}(M; {}^{b,c}\Omega^{1/2})$, $B_1 \in \widetilde{\Psi}_{b,c}^{-\infty,-\infty,\gamma}(M, {}^{b,c}\Omega^{1/2})$ and $B_2 \in \Psi^{-\infty,\gamma}(M, {}^{b,c}\Omega^{1/2})$, we have $AB_1, B_1 A \in \widetilde{\Psi}_{b,c}^{-\infty,-\infty,\gamma}(M, {}^{b,c}\Omega^{1/2})$ resp. $AB_2, B_2 A \in \Psi^{-\infty,\gamma}(M, {}^{b,c}\Omega^{1/2})$ and the b-indicial family satisfies*

$$I_b(AB_1)(z) = I_b(A)(z) I_b(B_1)(z) = I_b(B_1 A)(z)$$

for all $-1/2 - \gamma_{\mathrm{rb}} < \mathrm{Im}\, z < 1/2 + \gamma_{\mathrm{lb}}$. Moreover, the corresponding canonical bilinear maps

$$\Psi_{b,c}^{-\infty,-\infty}(M; {}^{b,c}\Omega^{1/2}) \times \widetilde{\Psi}_{b,c}^{-\infty,-\infty,\gamma}(M, {}^{b,c}\Omega^{1/2}) \longrightarrow \widetilde{\Psi}_{b,c}^{-\infty,-\infty,\gamma}(M, {}^{b,c}\Omega^{1/2}),$$

$$\widetilde{\Psi}_{b,c}^{-\infty,-\infty,\gamma}(M, {}^{b,c}\Omega^{1/2}) \times \Psi_{b,c}^{-\infty,-\infty}(M; {}^{b,c}\Omega^{1/2}) \longrightarrow \widetilde{\Psi}_{b,c}^{-\infty,-\infty,\gamma}(M, {}^{b,c}\Omega^{1/2}),$$

$$\Psi_{b,c}^{-\infty,-\infty}(M; {}^{b,c}\Omega^{1/2}) \times \Psi^{-\infty,\gamma}(M, {}^{b,c}\Omega^{1/2}) \longrightarrow \Psi^{-\infty,\gamma}(M, {}^{b,c}\Omega^{1/2}),$$

$$\Psi^{-\infty,\gamma}(M, {}^{b,c}\Omega^{1/2}) \times \Psi_{b,c}^{-\infty,-\infty}(M; {}^{b,c}\Omega^{1/2}) \longrightarrow \Psi^{-\infty,\gamma}(M, {}^{b,c}\Omega^{1/2})$$

are jointly continuous.

3.6. Fredholm theory for the b-c-calculus

For $s \in \mathbb{R}$, let us define *b-c-Sobolev spaces* by

$$H_{b,c}^s(M, {}^{b,c}\Omega^{1/2}) := \left\{ f \in \mathcal{C}^{-\infty}(M, {}^{b,c}\Omega^{1/2}) : \Psi_{b,c}^s(M; {}^{b,c}\Omega^{1/2}) f \subseteq L^2(M, {}^{b,c}\Omega^{1/2}) \right\}$$

and then the corresponding *weighted* version for $k_0, k_1 \in \mathbb{R}$ by

$$\varrho_0^{k_0} \varrho_1^{k_1} H_{b,c}^s(M, {}^{b,c}\Omega^{1/2}) := \left\{ f \in \mathcal{C}^{-\infty}(M, {}^{b,c}\Omega^{1/2}) : \varrho_0^{-k_0} \varrho_1^{-k_1} f \in H_{b,c}^s(M, {}^{b,c}\Omega^{1/2}) \right\}.$$

Then $A \in \Psi_{b,c}^{m,k}(M; {}^{b,c}\Omega^{1/2})$ extends to a bounded operator

(3.8) $\qquad A : \varrho_0^{k_0} \varrho_1^{k_1} H_{b,c}^s(M, {}^{b,c}\Omega^{1/2}) \to \varrho_0^{k_0} \varrho_1^{k_1 - k} H_{b,c}^{s-m}(M, {}^{b,c}\Omega^{1/2}).$

The Fredholm theory for b-c-operators on M is summarized in the following theorem [60, 68, 72].

THEOREM 3.6.1. *Let $\mathfrak{a}_0, \mathfrak{a}_1, s \in \mathbb{R}$, and $A \in \Psi_{b,c}^{m,k}(M; {}^{b,c}\Omega^{1/2})$ be arbitrary. Then the operator $A : \varrho_0^{\mathfrak{a}_0} \varrho_1^{\mathfrak{a}_1} H_{b,c}^s(M, {}^{b,c}\Omega^{1/2}) \to \varrho_0^{\mathfrak{a}_0} \varrho_1^{\mathfrak{a}_1 - k} H_{b,c}^{s-m}(M, {}^{b,c}\Omega^{1/2})$ is Fredholm if and only if*

- *${}^{b,c}\sigma^{(m,k)}(A)(\zeta) \neq 0$ for all $\zeta \in {}^{b,c}S^*M$,*
- *$I_b(A)(\xi - i\mathfrak{a}_0) \neq 0$ for all $\xi \in \mathbb{R}$, and*
- *$I_c^{(k)}(A)(\xi) \neq 0$ for all $\xi \in \mathbb{R}$.*

In that case, we find a weight $\gamma = (\gamma_{\mathrm{lb}}, \gamma_{\mathrm{rb}})$ satisfying
- $-\mathfrak{a}_0 \in (-1/2 - \gamma_{\mathrm{lb}}, 1/2 + \gamma_{\mathrm{rb}})$, and
- $I_b(A)(\xi + i\mu) \neq 0$ for all $\xi \in \mathbb{R}$ and all $\mu \in [-1/2 - \gamma_{\mathrm{lb}}, 1/2 + \gamma_{\mathrm{rb}}]$,

and $B = B_{\mathfrak{a}_0, \mathfrak{a}_1} \in \Psi_{b,c}^{-m,-k,\gamma}(M; {}^{b,c}\Omega^{1/2})$ such that $\mathrm{id} - BA$ resp. $\mathrm{id} - AB$ is the projection onto the kernel resp. the orthogonal complement to $R(A)$ in the Hilbert space $\varrho_0^{\mathfrak{a}_0}\varrho_1^{\mathfrak{a}_1} H_{b,c}^m(M, {}^{b,c}\Omega^{1/2})$ resp. $\varrho_0^{\mathfrak{a}_0}\varrho_1^{\mathfrak{a}_1-k} L^2(M, {}^{b,c}\Omega^{1/2})$.

In particular, the inverse of an invertible b-c-operators $A \in \Psi_{b,c}^{m,k}(M; {}^{b,c}\Omega^{1/2})$ belongs to the b-c-calculus with bounds $\Psi_{b,c}^{-m,-k,\gamma}(M; {}^{b,c}\Omega^{1/2})$ for an appropriate weight system γ as above.

For the proof of the Fredholm property of 0-pseudodifferential operators we need the following parameter dependent version of Theorem 3.6.1.

PROPOSITION 3.6.2. *Let P be a smooth manifold and $\mathcal{N} : P \to \Psi_{b,c}^0(M; {}^{b,c}\Omega^{1/2})$ be a smooth map such that $\mathcal{N}(p) : \varrho_0^{\mathfrak{a}_0}\varrho_1^{\mathfrak{a}_1} L^2(M, {}^{b,c}\Omega^{1/2}) \to \varrho_0^{\mathfrak{a}_0}\varrho_1^{\mathfrak{a}_1} L^2(M, {}^{b,c}\Omega^{1/2})$ is invertible for all $p \in P$, and let $\gamma = (\gamma_{\mathrm{lb}}, \gamma_{\mathrm{rb}})$ be such that $I_b(\mathcal{N}(p))(\xi + i\mu) \neq 0$ for all $\xi \in \mathbb{R}$, all $\mu \in [-1/2 - \gamma_{\mathrm{lb}}, 1/2 + \gamma_{\mathrm{rb}}]$ and all $p \in P$. Then there exist smooth functions $\mathcal{N}_0 : P \to \Psi_{b,c}^0(M; {}^{b,c}\Omega^{1/2})$, $\mathcal{N}_1 : P \to \widetilde{\Psi}_{b,c}^{-\infty,-\infty,\gamma}(M, {}^{b,c}\Omega^{1/2})$, and $\mathcal{N}_2 : P \to \Psi^{-\infty,\gamma'}(M, {}^{b,c}\Omega^{1/2})$ such that $\gamma' = \gamma + (1/2, 1/2)$ and*
$$\mathcal{N}(p)^{-1} = \mathcal{N}_0(p) + \mathcal{N}_1(p) + \mathcal{N}_2(p) \text{ for all } p \in P.$$

Proof: An inspection of the construction of the inverse of a b-pseudodifferential operator for instance in [68] shows that the three contribution depend smoothly on $p \in P$, see also [68, Lemma 6.23]. Note that the weight $\gamma = (\gamma_{\mathrm{lb}}, \gamma_{\mathrm{rb}})$ can be chosen uniformly in $p \in P$ by assumption. The additional cusp-part does not affect the construction because the cusp-calculus is inverse-closed anyway. $\square$

3.7. Invariance of the b-c-calculus under the $\mathbb{R}_+$-action

In what follows, we should mostly think of M as in (3.1) as the radial compactification of the half-axis $\overline{\mathbb{R}}_+$. The natural $\mathbb{R}_+$-action on $\overline{\mathbb{R}}_+$ extends by continuity to a smooth action of $\mathbb{R}_+$ on M. Explicitly, we have

$$(3.9) \quad \mathbb{R}_+ \times M \longrightarrow M : (t, z) \longmapsto t \cdot z := \begin{cases} \frac{2}{\pi} \arctan\left(t \tan(\frac{\pi}{2} z)\right) & , \quad 0 \leq z < 1 \\ 1 & , \quad z = 1 \end{cases}.$$

For $t \in \mathbb{R}_+$, the diffeomorphism $M \to M : z \mapsto t \cdot z$ extends via pull-back and duality to a map
$$Q_t : \mathcal{C}^{-\infty}(M, {}^{b,c}\Omega^{1/2}) \longrightarrow \mathcal{C}^{-\infty}(M, {}^{b,c}\Omega^{1/2})$$
satisfying
(3.10)
$$Q_t(\dot{\mathcal{C}}^\infty(M, {}^{b}\Omega^{1/2})) \subseteq \dot{\mathcal{C}}^\infty(M, {}^{b,c}\Omega^{1/2}) \text{ and } Q_t(\mathcal{C}^\infty(M, {}^{b}\Omega^{1/2})) \subseteq \mathcal{C}^\infty(M, {}^{b,c}\Omega^{1/2}).$$

It is often more convenient to consider first the non-compact case $\overline{\mathbb{R}}_+$, and then to check what happens at infinity, i.e. at $1 \in M$. Let us start with the action of Q_t on functions. Suppose $f \in \mathcal{C}^\infty(M, {}^{b,c}\Omega^{1/2})$, then we have $f = \widehat{f}(x) \left|\frac{dx}{x}\right|^{\frac{1}{2}}$ for some $\widehat{f} \in S_{cl}^{1/2}(; \mathbb{R})|_{\overline{\mathbb{R}}_+}$, and the action of Q_t is given by

$$(3.11) \quad (Q_t f)(x) = \widehat{f}(tx) \left|\frac{dx}{x}\right|^{\frac{1}{2}} \text{ for } x \in \overline{\mathbb{R}}_+,$$

Similarly, for the lifted Schwartz kernel $\kappa_H \in \mathcal{C}^{-\infty}(M_{b,c}^2, KD_{b,c}^{1/2})$ of a continuous linear operator $H : \dot{\mathcal{C}}^\infty(M, {}^{b,c}\Omega^{1/2}) \to \mathcal{C}^{-\infty}(M, {}^{b,c}\Omega^{1/2})$, we first consider the behavior near the b-part via $\beta_b^2 : [-1,1] \times \overline{\mathbb{R}}_+ \to \overline{\mathbb{R}}_+ \times \overline{\mathbb{R}}_+$, and then deal with the behavior as $r \to \infty$. In fact, we have $\kappa_H = \widehat{\kappa}_H(\tau, r) \left|\frac{dr}{r} d\tau\right|^{\frac{1}{2}}$ for some $\widehat{\kappa}_H \in \mathcal{C}^{-\infty}([-1,1] \times \overline{\mathbb{R}}_+)$, and the lifted Schwartz kernel $\kappa_{H_t} = \widehat{\kappa}_{H_t}(\tau, r) \left|\frac{dr}{r} d\tau\right|^{\frac{1}{2}}$ of $H_t := Q_t H Q_{t^{-1}}$ satisfies

$$(3.12) \qquad \kappa_{H_t} = \widehat{\kappa}_H(\tau, tr) \left|\frac{dr}{r} d\tau\right|^{\frac{1}{2}}.$$

The b-c-calculus is invariant under conjugation with Q_t, more precisely, we have:

PROPOSITION 3.7.1. *Let $t > 0$ and $A \in \Psi_{b,c}^{m,k,\gamma}(M; {}^{b,c}\Omega^{1/2})$, be arbitrary. Then we have $A_t := Q_t A Q_{t^{-1}} \in \Psi_{b,c}^{m,k,\gamma}(M; {}^{b,c}\Omega^{1/2})$ and*

$$\begin{aligned}
{}^{b,c}\sigma^{(0,0)}(A_t)(z, \pm 1) &= {}^{b,c}\sigma^{(0,0)}(A)(t \cdot z, \pm 1) \text{ provided } m = k = 0, \\
I_b(A_t)(z) &= I_b(A)(z) \text{ for } z \in \mathbb{C} \text{ with } -\frac{1}{2} - \gamma_{\mathrm{rb}} < \operatorname{Im} z < \frac{1}{2} + \gamma_{\mathrm{lb}}, \\
I_c^{(k)}(A_t)(\xi) &= t^k I_c^{(k)}(A)(\xi/t) \text{ for } \xi \in \mathbb{R}_\xi.
\end{aligned}$$

Proof: A straightforward computation using (3.12) gives $A_t \in \Psi_{b,c}^{m,k,\gamma}(M; {}^{b,c}\Omega^{1/2})$. As a byproduct we obtain the formulas for the homogeneous principal symbol, and the b- resp. c-indicial families of A_t. □

3.8. C^*-algebras of b-c-operators

Throughout this section fix two real numbers $\mathfrak{a}_0, \mathfrak{a}_1 \in \mathbb{R}$, and let us write $\varrho_0^{\mathfrak{a}_0} \varrho_1^{\mathfrak{a}_1} L^2(M, {}^{b,c}\Omega^{1/2})$ instead of $\varrho_0^{\mathfrak{a}_0} \varrho_1^{\mathfrak{a}_1} H_{b,c}^0(M, {}^{b,c}\Omega^{1/2})$. Note that the Hilbert space structure of $\varrho_0^{\mathfrak{a}_0} \varrho_1^{\mathfrak{a}_1} L^2(M, {}^{b,c}\Omega^{1/2})$, and hence also the C^*-algebra structure of the algebra of all bounded operators $\mathcal{L}(\varrho_0^{\mathfrak{a}_0} \varrho_1^{\mathfrak{a}_1} L^2(M, {}^{b,c}\Omega^{1/2}))$ depends on the choice of the defining functions ϱ_0 resp. ϱ_1 as long as $\mathfrak{a}_0 \neq 0$ resp. $\mathfrak{a}_1 \neq 0$.

By (3.5), (3.8), and Theorem 3.5.2, $\Psi_{b,c}^0(M; {}^{b,c}\Omega^{1/2})$ can be realized as a symmetric subalgebra of $\mathcal{L}(\varrho_0^{\mathfrak{a}_0} \varrho_1^{\mathfrak{a}_1} L^2(M, {}^{b,c}\Omega^{1/2}))$, by Theorem 3.6.1 it is even a subalgebra with a $\mathcal{K}(\varrho_0^{\mathfrak{a}_0} \varrho_1^{\mathfrak{a}_1} L^2(M, {}^{b,c}\Omega^{1/2}))$-symbolic structure in the sense of [40]. Recall that $\mathcal{K}(\varrho_0^{\mathfrak{a}_0} \varrho_1^{\mathfrak{a}_1} L^2(M, {}^{b,c}\Omega^{1/2}))$ stands for the ideal of compact operators on the space $\varrho_0^{\mathfrak{a}_0} \varrho_1^{\mathfrak{a}_1} L^2(M, {}^{b,c}\Omega^{1/2})$. The symbol morphism is given by

$$\begin{aligned}
\tau_{b,c} : \Psi_{b,c}^0(M; {}^{b,c}\Omega^{1/2}) &\longrightarrow \mathcal{C}({}^{b,c}S^*M) \oplus \mathcal{B}(;\Gamma_{\mathfrak{a}_0}) \oplus \mathcal{B}(;\mathbb{R}) =: Q : \\
A &\longmapsto \left({}^{b,c}\sigma^{(0,0)}(A), I_b(A)|_{\Gamma_{\mathfrak{a}_0}}, I_c(A)\right)
\end{aligned}$$

where $\Gamma_{\mathfrak{a}_0} = \{z \in \mathbb{C} : \operatorname{Im} z = -\mathfrak{a}_0\} \cong \mathbb{R}$, and $\mathcal{B}(;\Gamma_{\mathfrak{a}_0})$ resp. $\mathcal{B}(;\mathbb{R})$ stands for the closure of $M_\mathcal{O}^0|_{\Gamma_{\mathfrak{a}_0}}$ resp. $S_{cl}^0(;\mathbb{R})$ in the C^*-algebra $\mathcal{C}_b(\Gamma_{\mathfrak{a}_0})$ resp. $\mathcal{C}_b(\mathbb{R})$. Note that by [52, Corollary 3.3] – see also [64, 92] – the C^*-algebras $\mathcal{B}(;\Gamma_{\mathfrak{a}_0})$ and $\mathcal{B}(;\mathbb{R})$ are naturally isomorphic. They are completely determined by the exactness of the following sequence of C^*-algebras

$$(3.13) \qquad 0 \longrightarrow \mathcal{C}_0(\mathbb{R}) \longrightarrow \mathcal{B}(;\mathbb{R}) \xrightarrow{\sigma_\mathcal{B}^{(0)}} \mathbb{C} \oplus \mathbb{C} \longrightarrow 0,$$

where $\sigma_\mathcal{B}^{(0)} : B(;\mathbb{R}) \to \mathcal{C}(\{\pm 1\}) = \mathbb{C} \oplus \mathbb{C}$ is the extension by continuity of the homogeneous principal part $\sigma^{(0)} : S_{cl}^0(;\mathbb{R}) \to \mathcal{C}^\infty(\{\pm 1\}) = \mathbb{C} \oplus \mathbb{C}$ [52]. Therefore,

$\mathcal{B}(;\mathbb{R}) \cong \mathcal{C}([-1,1])$ in an obvious way. However, we keep the notation $\mathcal{B}(;\mathbb{R})$ for systematic reasons – see [**39, 41**].

Let $\mathcal{B}_{b,c}^{(\mathfrak{a}_0,\mathfrak{a}_1)}(M,{}^{b,c}\Omega^{1/2})$ be the C^*-algebra generated by $\Psi_{b,c}^0(M;{}^{b,c}\Omega^{1/2})$ in the C^*-algebra $\mathcal{L}(\varrho_0^{\mathfrak{a}_0}\varrho_1^{\mathfrak{a}_1}L^2(M,{}^{b,c}\Omega^{1/2}))$. Since the space of operators with Schwartz kernels in $\dot{\mathcal{C}}^\infty(M^2,{}^{b,c}\Omega^{1/2}\boxtimes^{b,c}\Omega^{1/2})$ is dense in $\mathcal{K}(\varrho_0^{\mathfrak{a}_0}\varrho_1^{\mathfrak{a}_1}L^2(M,{}^{b,c}\Omega^{1/2}))$ and belongs to $\Psi_{b,c}^{-\infty}(M;{}^{b,c}\Omega^{1/2})$, we have

$$\mathcal{K}(\varrho_0^{\mathfrak{a}_0}\varrho_1^{\mathfrak{a}_1}L^2(M,{}^{b,c}\Omega^{1/2})) \subseteq \mathcal{B}_{b,c}^{(\mathfrak{a}_0,\mathfrak{a}_1)}(M,{}^{b,c}\Omega^{1/2}).$$

Moreover, by [**40**, Proposition 2.31], the symbol morphism $\tau_{b,c}$ extends by continuity to the algebra $\mathcal{B}_{b,c}^{(\mathfrak{a}_0,\mathfrak{a}_1)}(M,{}^{b,c}\Omega^{1/2})$ and yields the following short exact sequence of C^*-algebras
(3.14)
$$0 \longrightarrow \mathcal{K}(\varrho_0^{\mathfrak{a}_0}\varrho_1^{\mathfrak{a}_1}L^2(M,{}^{b,c}\Omega^{1/2})) \longrightarrow \mathcal{B}_{b,c}^{(\mathfrak{a}_0,\mathfrak{a}_1)}(M,{}^{b,c}\Omega^{1/2}) \xrightarrow{\tau_{b,c}} Q_{b,c}^{(\mathfrak{a}_0,\mathfrak{a}_1)} \longrightarrow 0,$$

where the C^*-algebra of *joint b-c-symbols* $Q_{b,c}^{(\mathfrak{a}_0,\mathfrak{a}_1)} := R(\tau_{b,c})$ consists of all triples $(f_0, f_1, f_2) \in \mathcal{C}({}^{b,c}S^*M) \oplus \mathcal{B}(;\Gamma_{\mathfrak{a}_0}) \oplus \mathcal{B}(;\mathbb{R})$ satisfying

(3.15) $\quad\quad\quad\quad\quad\quad \sigma_{\mathcal{B}}^{(0)}(f_1)(\pm 1) = f_0(0,\pm 1)$ and

(3.16) $\quad\quad\quad\quad\quad\quad \sigma_{\mathcal{B}}^{(0)}(f_2)(\pm 1) = f_0(1,\pm 1)$.

Indeed, by Theorem 3.5.2 we know $\tau_{b,c}(\Psi_{b,c}^0(M;{}^{b,c}\Omega^{1/2})) \subseteq Q_{b,c}^{(\mathfrak{a}_0,\mathfrak{a}_1)}$. On the other hand, the set Q_ψ of all $(f_0, f_1, f_2) \in \mathcal{C}^\infty({}^{b,c}S^*M) \oplus M_{\mathcal{O}}^0|_{\Gamma_{\mathfrak{a}_0}} \oplus S_{cl}^0(;\mathbb{R})$ such that (3.15) and (3.16) holds, is dense in $Q_{b,c}^{(\mathfrak{a}_0,\mathfrak{a}_1)}$ and satisfies $Q_\psi \subseteq \tau(\Psi_{b,c}^0(M;{}^{b,c}\Omega^{1/2}))$.

Let us denote the composition of $\tau_{b,c} : \mathcal{B}_{b,c}^{(\mathfrak{a}_0,\mathfrak{a}_1)}(M,{}^{b,c}\Omega^{1/2}) \to Q_{b,c}^{(\mathfrak{a}_0,\mathfrak{a}_1)}$ with the projection onto the first resp. second resp. third component by ${}^{b,c}\sigma^{(0,0)}$ resp. $I_b^{\mathfrak{a}_0}$ resp. I_c.

Thus, $\mathcal{B}_{b,c}^{(\mathfrak{a}_0,\mathfrak{a}_1)}(M,{}^{b,c}\Omega^{1/2})$ is an essentially commutative C^*-algebra, i.e. a C^*-algebra that commutes up to the ideal of compact operators. Finally, let us give a decomposition of $Q_{b,c}^{(\mathfrak{a}_0,\mathfrak{a}_1)}$ that resolves the compatibility conditions (3.15) and (3.16).

PROPOSITION 3.8.1. *Let* $\mathcal{J}_0 := \{(f_0, f_1, f_2) \in Q_{b,c}^{(\mathfrak{a}_0,\mathfrak{a}_1)} : f_0 = 0\}$. *Then* $\mathcal{J}_0$ *is a closed ideal in* $Q_{b,c}^{(\mathfrak{a}_0,\mathfrak{a}_1)}$ *and we have*

$$Q_{b,c}^{(\mathfrak{a}_0,\mathfrak{a}_1)}/\mathcal{J}_0 \cong \mathcal{C}({}^{b,c}S^*M) \cong \mathcal{C}(M) \oplus \mathcal{C}(M) \text{ and}$$
$$\mathcal{J}_0 \cong \mathcal{C}_0(\mathbb{R}) \oplus \mathcal{C}_0(\mathbb{R}).$$

A description of the Jacobson topology on the spectra of $\mathcal{B}_{b,c}^{(\mathfrak{a}_0,\mathfrak{a}_1)}(M,{}^{b,c}\Omega^{1/2})$ resp. $Q_{b,c}^{(\mathfrak{a}_0,\mathfrak{a}_1)}$ can easily be deduced from this representation. In fact, it is a special case of [**41**, Theorem 4.1.6, Corollary 4.1.7].

3.9. General bundles

Exactly as the 0-calculus, the *b-c*-calculus on M extends to pseudodifferential operators acting between sections of finite-dimensional vector bundles $E, F \to M$ over M by defining $\Psi_{b,c}^{m,k,\gamma}(M;E,F)$ to be the space
(3.17)
$$\Psi_{b,c}^{m,k,\gamma}(M;{}^{b,c}\Omega^{1/2}) \otimes_{\mathcal{C}^\infty(M_{b,c}^2)} \mathcal{C}^\infty(M_{b,c}^2, (\beta_{b,c}^2)^*(\text{Hom}(E \otimes {}^{b,c}\Omega^{-1/2}, F \otimes {}^{b,c}\Omega^{-1/2})).$$

As one might expect, most of the properties of b-c-pseudodifferential operators remain true with the obvious modifications. We leave the details to the reader.

CHAPTER 4

The reduced normal operator

In this section we are going to associate to each 0-pseudodifferential operator a family of b-c-operators acting on the interval $M = [0, 1]$. The invertibility of this family together with the invertibility of the homogeneous principal 0-symbol then is equivalent to the Fredholm property of 0-operators.

First, we define the reduced normal operator using local coordinates as an element of the space $\mathcal{C}^{-\infty}([-1, 1] \times \overline{\mathbb{R}}_+, r^{-1/2}\Omega^{1/2})$, then, we describe how this element depends on the choice of local coordinates. In a second step we characterize the behavior of the reduced normal operator for $r \to 0$ and $r \to \infty$, showing in particular that the reduced normal operator extends to an element in $\mathcal{C}^{-\infty}(M^2_{b,c}, {}^{b,c}\Omega^{1/2})$ belonging to a b-c-pseudodifferential operator on M.

4.1. Definition of the reduced normal operator

Choose a defining function $\varrho_N : X \to \overline{\mathbb{R}}_+$ for the boundary ∂X of X. Note that the choice of ϱ_N uniquely determines a trivialization of the positive normal bundle

$$(4.1) \quad \Phi_{\varrho_N} : N^+\partial X := T^+X|_{\partial X}/T\partial X \xrightarrow{\cong} \overline{\mathbb{R}}_+ \times \partial X : [v_q] \longmapsto (d\varrho_N|_q(v_q), q)$$

of the boundary. Furthermore, let ν be a *normal fibration* to the boundary, i.e. $\nu : \Gamma \to X$ is a diffeomorphism of an open neighborhood $\partial X \subseteq \Gamma \subseteq N^+\partial X$ of the zero-section ∂X in $N^+\partial X$ onto $\nu(\Gamma) \subseteq X$ such that $\nu|_{\partial X} : \partial X \to X$ is the inclusion, and ν induces the identity $\nu_* : N^+\partial X \to N^+\partial X$. The existence of normal fibrations is discussed for instance in [9] or [62, Proposition 2.10.1]. Finally, choose local coordinates $\chi : \partial X \supseteq V^{\partial X} \to \mathbb{R}^{n-1}_y$ on the boundary ∂X. Thus, we obtain for each $q \in V^{\partial X}$ local coordinates

$$(4.2) \quad (x, y) : X \supseteq V^X := \nu(\Phi_{\varrho_N}^{-1}(\overline{\mathbb{R}}_+ \times V^{\partial X})) \longrightarrow \overline{\mathbb{R}}_+ \times \mathbb{R}^{n-1}_y$$

near q. As in (1.8), the coordinates from (4.2) lead to a system of projective coordinates (τ, U, r, y') near the interior of the extended 0-front face $\mathrm{ff}^{0,e}$, and the lifted Schwartz kernel κ_A of a 0-operator $A \in \Psi_0^{m,0,\gamma}(X; {}^0\Omega^{1/2})$ can locally be written near $\mathrm{ff}^{0,e}$ as $\kappa_A = \widehat{\kappa}_A(\tau, U, r, y')\left|d\tau\, dU\, \frac{dr}{r^n}\, dy'\right|^{\frac{1}{2}}$ with

$$(4.3) \quad \widehat{\kappa}_A \in \mathcal{C}^{-\infty}([-1,1])\widehat{\otimes}_\pi \mathscr{S}'(\mathbb{R}^{n-1}_U)\widehat{\otimes}_\pi \mathcal{C}^{-\infty}(\overline{\mathbb{R}}_+)\widehat{\otimes}_\pi \mathcal{C}^{-\infty}(\mathbb{R}^{n-1}_{y'}).$$

For simplicity, we fix $q \in \partial X$, and write $(x, y)(q) = (0, \bar{y})$ for some $\bar{y} \in \mathbb{R}^{n-1}_y$. For $\eta \in \mathbb{R}^{n-1}_\eta \setminus \{0\}$ and $\bar{y} \in \mathbb{R}^{n-1}_y$, we define the *reduced normal operator* $\mathcal{N}^{\nu,\chi}_{\varrho_N}(A)$ of

$A \in \Psi_0^{m,0,\gamma}(X; {}^0\Omega^{1/2})$ at $(0,\bar{y}) = q \in \partial X$ to be the distributional density

$$(4.4) \quad \mathcal{N}_{\varrho N}^{\nu,\chi}(A)(\bar{y},\eta;\tau,r) := \mathcal{F}_{\varrho N}^{\nu,\chi}(A)(\bar{y},r\eta;\tau) \left|\frac{dr}{r}d\tau\right|^{\frac{1}{2}}$$

$$\in \mathcal{C}^{-\infty}([-1,1], \Omega^{1/2}) \hat{\otimes}_\pi \mathcal{C}^{-\infty}(\overline{\mathbb{R}}_+, {}^b\Omega^{1/2}) \text{ where}$$

$$(4.5) \quad \mathcal{F}_{\varrho N}^{\nu,\chi}(A)(\bar{y},\eta;\tau) := \int_{\mathbb{R}_U^{n-1}} e^{-iU\eta} \widehat{\kappa}_A(\tau,U,0,\bar{y}) dU \in \mathcal{C}^{-\infty}([-1,1]).$$

Since $\widehat{\kappa}_A$ is conormal at $U = 0$ and symbolic of order at most γ_F by (2.14), the partial Fourier transform of $\widehat{\kappa}_A$ with respect to U is well-defined and smooth for $\eta \neq 0$. Note that for $\gamma_F > n - 1$, (4.5) continues to make sense even for $\eta = 0$. The variables τ and r belong to the new space $[-1,1] \times \overline{\mathbb{R}}_+$ that is obtained by blowing up the corner $(0,0)$ in $\overline{\mathbb{R}}_+ \times \overline{\mathbb{R}}_+$ as in (3.2).

4.2. Coordinate invariance of the reduced normal operator

Let $\nu, \tilde{\nu}$ be two normal fibrations of ∂X, and $\chi, \tilde{\chi}$ be two local coordinate systems of ∂X near q. For the change of coordinates (x,y) resp. $(\tilde{x}, \tilde{y})$ on X given by ν, χ resp. $\tilde{\nu}, \tilde{\chi}$ via (4.2), we can then assume that we have after appropriate restrictions

$$(4.6) \quad (\tilde{x}, \tilde{y}) = \varphi(x,y) = (x\varphi_0(x,y), \varphi_1(x,y))$$

where $\tilde{y} = \varphi_1(0,y) = \tilde{\chi} \circ \chi^{-1}(y)$ is the change of coordinates on ∂X, and we have $\varphi_0(0,y) = 1$ by the normalization property of normal fibrations. The inverse $\psi = \varphi^{-1}$ has of course the same properties.

Let us denote by $(\tilde{\tau}, \tilde{U}, \tilde{r}, \tilde{y}')$ the coordinates near ff^{0,e} induced by lifting the coordinates $(\tilde{x}, \tilde{y}, \tilde{x}', \tilde{y}')$ from X^2 to $X_{0,e}^2$ as in (1.8). The corresponding lifted Schwartz kernels in (4.3) are then related by
$$(4.7) \quad \widehat{\tilde{\kappa}}_A(\tilde{\tau}, \tilde{U}, 0, \tilde{y}') = \widehat{\kappa}_A\left(\tilde{\tau}, \tilde{\tau}\partial_1\psi_1(0,\tilde{y}') + \partial_2\psi_1(0,\tilde{y}')\tilde{U}, 0, \psi_1(0,\tilde{y}')\right) |\partial_2\psi_1(0,\tilde{y}')|.$$

Inserting (4.7) into the defining equations (4.4) and (4.5) for the reduced normal operator and substituting

$$U = \tau \partial_1 \psi_1(0,\tilde{y}) + \partial_2\psi_1(0,\tilde{y})\tilde{U}$$

leads finally to the identities

$$(4.8) \quad F_{\varrho N}^{\tilde{\nu},\tilde{\chi}}(A)(\tilde{y},\eta;\tau) = e^{-i\tau(\partial_1\psi_1)(0,\tilde{y})((\partial_2\psi_1(0,\tilde{y}))^{-1})^*\tilde{\eta}} \times$$
$$\times \mathcal{F}_{\varrho N}^{\nu,\chi}(A)(\psi_1(0,\tilde{y}), ((\partial_2\psi_1(0,\tilde{y}))^{-1})^*\tilde{\eta}) \text{ and}$$

$$(4.9) \quad \mathcal{N}_{\varrho N}^{\tilde{\nu},\tilde{\chi}}(A)(\tilde{y},\tilde{\eta};\tau,r) = e^{-ir\tau(\partial_1\psi_1)(0,\tilde{y})((\partial_2\psi_1(0,\tilde{y}))^{-1})^*\tilde{\eta}} \times$$
$$\times \mathcal{N}_{\varrho N}^{\nu,\chi}(A)(\psi_1(0,\tilde{y}), ((\partial_2\psi_1(0,\tilde{y}))^{-1})^*\tilde{\eta}).$$

This shows that for $\nu = \tilde{\nu}$, i.e. $(\partial_1\psi_1)(0,\tilde{y}) = 0$, the reduced normal operator as well as the function $\mathcal{F}_{\varrho N}^{\nu,\chi}$ are well-defined as functions on the cotangent bundle $T^*\partial X \setminus \{0\}$ of ∂X, whereas for arbitrary ν and $\tilde{\nu}$ the reduced normal operator, for instance, corresponds to a section of the following bundle over $T^*\partial X \setminus \{0\}$. Let E be the set of all tuples $(\nu, \chi, y, \eta, \mathcal{N}(y,\eta))$ where ν is a normal fibration of the boundary ∂X, χ is a local coordinate system on the boundary ∂X, $y \in \mathbb{R}_y^{n-1}$, $\eta \in \mathbb{R}_\eta^{n-1} \setminus \{0\}$, and $\mathcal{N}(y,\eta) \in \mathcal{C}^{-\infty}([-1,1], \Omega^{1/2}) \hat{\otimes}_\pi \mathcal{C}^{-\infty}(\overline{\mathbb{R}}_+, {}^b\Omega^{1/2})$. We call two such tuples $(\nu, \chi, y, \eta, \mathcal{N}(y, \eta))$ and $(\widetilde{\nu}, \widetilde{\chi}, \widetilde{y}, \widetilde{\eta}, \widetilde{\mathcal{N}}(y, \eta))$ equivalent provided the domains of χ and $\widetilde{\chi}$ have a non-trivial intersection, and for a change of coordinates $(\widetilde{x}, \widetilde{y}) = \varphi(x, y)$ as in (4.6) we have $y = \psi_1(0, \widetilde{y})$, $\eta = ((\partial_2 \psi_1(0, \widetilde{y}))^{-1})^* \widetilde{\eta}$, and $\widetilde{\mathcal{N}}(\widetilde{y}, \widetilde{\eta}) = e^{-ir\tau (\partial_1 \psi_1)(0, \widetilde{y})} ((\partial_2 \psi_1(0, \widetilde{y}))^{-1})^* \widetilde{\eta} \mathcal{N}(\psi_1(0, \widetilde{y}), ((\partial_2 \psi_1(0, \widetilde{y}))^{-1})^* \widetilde{\eta})$. The set $\mathcal{E}$ of equivalence classes is equipped with a projection map $\pi : \mathcal{E} \to T^* \partial X \setminus \{0\}$ that is onto, and the choice of a normal fibration clearly trivializes the bundle $\mathcal{E}$. As desired, any $A \in \Psi_0^{m,0,\gamma}(X; {}^0\Omega^{1/2})$ induces a section $\mathcal{N}_{\varrho_N}(A) : T^* \partial X \setminus \{0\} \to \mathcal{E}$, that is still called the *reduced normal operator of A*. The dependence of the reduced normal operator on the choice of the defining function ϱ_N is even a little bit more subtle involving rescalings of the bundle $T^* \partial X \setminus \{0\}$, and we do not want to enter into this. Instead, we consider the defining function ϱ_N to be fixed once and for all; for the sake of simplicity, we mostly fix a normal fibration ν of the boundary as well, and consider the bundle $\mathcal{E}$ trivialized by this choice, i.e. for $A \in \Psi_0^{m,0,\gamma}(X; {}^0\Omega^{1/2})$, we have maps

$$\mathcal{F}_{\varrho_N}^\nu(A) : T^* \partial X \setminus \{0\} \longrightarrow \mathcal{C}^{-\infty}([-1, 1]) \text{ and}$$
$$\mathcal{N}_{\varrho_N}^\nu(A) : T^* \partial X \setminus \{0\} \longrightarrow \mathcal{C}^{-\infty}([-1, 1], \Omega^{1/2}) \widehat{\otimes}_\pi \mathcal{C}^{-\infty}(\mathbb{R}_+, {}^b\Omega^{1/2}).$$

Let us briefly explain this coordinate invariance from a slightly more global point of view. The choice of a boundary defining function ϱ_N and a normal fibration ν to the boundary yields in particular a collar neighborhood $X \supseteq V \to \partial X \times \overline{\mathbb{R}}_+$ of the boundary. This leads to a splitting of the sequence (1.1), hence to an isomorphism

$$(4.10) \qquad {}^0TX|_{\partial X} \cong {}^0T \partial X \oplus {}^0TX|_{\partial X}/{}^0T \partial X;$$

on the other hand, we know that the interior of the extended 0-front-face $\mathrm{ff}^{0,e}$ can be identified with ${}^0TX|_{\partial X}$, thus, the function $\mathcal{F}_{\varrho_N}^\nu(A)$ is simply the partial Fourier transform of the restriction of the lifted Schwartz kernel κ_A to the extended 0-front face $\mathrm{ff}^{0,e}$ along the direction ${}^0T \partial X$; this leads to a well-defined function $\mathcal{F}_{\varrho_N}^\nu(A)$ on ${}^0T^* \partial X \cong T^* \partial X$ by the identification (1.2). However, in the sequel we mostly use the local representation (4.4) for the reduced normal operator because it is more convenient for concrete computations.

4.3. Scale invariance of the reduced normal operator

An immediate consequence of the definition (4.4) of the reduced normal operator is that r and $|\eta|$ are non-trivially related – indeed, we have for $t > 0$

$$(4.11) \qquad \mathcal{N}_{\varrho_N}^{\nu,\chi}(A)(y, t\eta; \tau, r) = \mathcal{N}_{\varrho_N}^{\nu,\chi}(y, \eta; \tau, tr) = \mathcal{F}_{\varrho_N}^{\nu,\chi}(A)(y, rt\eta; \tau) \left| \frac{dr}{r} d\tau \right|^{\frac{1}{2}}.$$

This identity can be understood in the following way.

COROLLARY 4.3.1. *For $A \in \Psi_0^{m,0,\gamma}(X; {}^0\Omega^{1/2})$, $t > 0$, and $\eta \in T^* \partial X \setminus \{0\}$, the reduced normal operator satisfies*

$$(4.12) \quad \mathcal{N}_{\varrho_N}^\nu(A)(t\eta) = Q_t \mathcal{N}_{\varrho_N}^\nu(A)(\eta) Q_{t^{-1}} : \dot{\mathcal{C}}_c^\infty(\overline{\mathbb{R}}_+, {}^b\Omega^{1/2}) \longrightarrow \mathcal{C}^{-\infty}(\overline{\mathbb{R}}_+, {}^b\Omega^{1/2}).$$

Proof: This follows immediately from the identity (3.12) for the lifted Schwartz kernel of $Q_t \mathcal{N}_{\varrho_N}^\nu(A)(\eta) Q_{t^{-1}}$ and (4.11). □

4.4. Characterization of the reduced normal operator

We are now going to characterize the indicial function and the reduced normal operator of 0-pseudodifferential operators. The reduced normal operator extends, in fact, to a family of b-c-operators on the interval $M = [0, 1]$; the indicial function corresponds to the indicial operator of the reduced normal operator at the b-end. Let $\pi : T^*\partial X \to \partial X \hookrightarrow T^*\partial X$ be the composition of the canonical projection with the inclusion of ∂X as the zero-section. For simplicity, we fix a normal fibration ν of the boundary and consider the following three cases separately:

- $A \in \Psi_0^{-\infty}(X; {}^0\Omega^{1/2})$,
- $A \in \Psi_0^m(X; {}^0\Omega^{1/2})$, and
- $A \in \Psi_0^{-\infty,0,\gamma}(X; {}^0\Omega^{1/2})$

The case $\Psi_0^{-\infty}(X; {}^0\Omega^{1/2})$.

PROPOSITION 4.4.1. *For a 0-pseudodifferential operator $A \in \Psi_0^{-\infty}(X; {}^0\Omega^{1/2})$, we have $\mathcal{F}_{\varrho_N}^\nu(A) \in \mathscr{S}(T^*\partial X) \widehat{\otimes}_\pi \dot{\mathcal{C}}^\infty([-1,1])$, and for any $\eta \in T^*\partial X \setminus \{0\}$, the operator $\mathcal{N}_{\varrho_N}^\nu(A)(\eta)$ extends uniquely to an element in $\Psi_{b,c}^{-\infty,-\infty}(M; {}^{b,c}\Omega^{1/2})$ such that $\mathcal{N}_{\varrho_N}^\nu(A) \in \mathcal{C}^\infty(T^*\partial X \setminus \{0\}, \Psi_{b,c}^{-\infty,-\infty}(M; {}^{b,c}\Omega^{1/2}))$ and*

$$(4.13) \quad I_c^{(k)}(\mathcal{N}_{\varrho_N}^\nu(A)(\eta)) = 0 \text{ for all } k \in \mathbb{Z}, \text{ and}$$

$$(4.14) \quad I_b\left(\mathcal{N}_{\varrho_N}^\nu(A)(\eta)\right)(z) = \int_{-1}^1 \left(\frac{1+\tau}{1-\tau}\right)^{-iz} \mathcal{F}_{\varrho_N}^\nu(A)(\pi(\eta); \tau) \frac{d\tau}{(1-\tau^2)^{1/2}}$$

$$= I_A(z - i\frac{n-1}{2})(\pi(\eta)).$$

If $\mathfrak{N}$ denotes the space of all pairs $(\mathcal{N}, \mathcal{F})$ with $\mathcal{F} \in \mathscr{S}(T^*\partial X) \widehat{\otimes}_\pi \dot{\mathcal{C}}^\infty([-1,1])$ and $\mathcal{N} \in \mathcal{C}^\infty(T^*\partial X \setminus \{0\}) \widehat{\otimes}_\pi \dot{\mathcal{C}}^\infty([-1,1], \Omega^{1/2}) \widehat{\otimes}_\pi \mathcal{C}^\infty(\overline{\mathbb{R}_+}, {}^b\Omega^{1/2})$ satisfying:

(a) $\mathcal{N}(\eta; \tau, r) = \mathcal{F}(r\eta; \tau) \left|d\tau \frac{dr}{r}\right|^{\frac{1}{2}}$ *for all $\eta \in T^*\partial X$, all $\tau \in [-1,1]$, and all $r \in \overline{\mathbb{R}_+}$.*

(b) *For each $\eta \in T^*\partial X \setminus \{0\}$, $\mathcal{N}(\eta)$ extends to $\mathcal{N}(\eta) \in \Psi_{b,c}^{-\infty,-\infty}(M; {}^{b,c}\Omega^{1/2})$, and we have $\mathcal{N}|_{T^*\partial X \setminus \{0\}} \in \mathcal{C}^\infty(T^*\partial X \setminus \{0\}, \Psi_{b,c}^{-\infty,-\infty}(M; {}^{b,c}\Omega^{1/2}))$.*

(c) $I_b\left(\mathcal{N}(\eta)\right)(z) = \int_{-1}^1 \left(\frac{1+\tau}{1-\tau}\right)^{-iz} \mathcal{F}(\pi(\eta); \tau) \frac{d\tau}{(1-\tau^2)^{1/2}}$ *for all $\eta \in T^*\partial X \setminus \{0\}$ and all $z \in \mathbb{C}$.*

Then we have $\mathfrak{N} = \{(\mathcal{N}_{\varrho_N}^\nu(A), \mathcal{F}_{\varrho_N}^\nu(A)) : A \in \Psi_0^{-\infty}(X; {}^0\Omega^{1/2})\}$.

Proof: We use the local representations (4.4) and (4.5) for the reduced normal operator and the map $\mathcal{F}_{\varrho_N}^{\nu,X}(A)$. Because of $A \in \Psi_0^{-\infty}(X; {}^0\Omega^{1/2})$ we have $\widehat{\kappa}_A \in \dot{\mathcal{C}}^\infty([-1,1]) \widehat{\otimes}_\pi \mathscr{S}(\mathbb{R}_U^{n-1}) \widehat{\otimes}_\pi \mathcal{C}_c^\infty(\overline{\mathbb{R}_+}) \widehat{\otimes}_\pi \mathcal{C}_c^\infty(\mathbb{R}_{y'}^{n-1})$; thus, after Fourier transform with respect to U, we have $\mathcal{F}_{\varrho_N}^\nu(A) \in \mathscr{S}(T^*\partial X) \widehat{\otimes}_\pi \dot{\mathcal{C}}^\infty([-1,1])$, hence the map $\mathcal{N}_{\varrho_N}^\nu(A) : T^*\partial X \to \dot{\mathcal{C}}^\infty([-1,1], \Omega^{1/2}) \widehat{\otimes}_\pi \mathcal{C}^\infty(\overline{\mathbb{R}_+}, {}^b\Omega^{1/2})$ is smooth, and for fixed $\eta \in T^*\partial X \setminus \{0\}$, the kernel $\widehat{\mathcal{N}_{\varrho_N}^\nu(A)}(\eta)(\tau, r)$ is rapidly decreasing as $r \to \infty$, and vanishes with all derivatives at $\tau = \pm 1$. This shows that $\mathcal{N}_{\varrho_N}^\nu(A)(\eta)$ extends to an operator in $\Psi_{b,c}^{-\infty,-\infty}(M; {}^{b,c}\Omega^{1/2})$ satisfying (4.13) and depending smoothly on $\eta \in T^*\partial X \setminus \{0\}$; (4.14) is just a combination of (2.15), (3.6), and (4.4). Since (a) follows from the very definition, we obtain $\mathcal{N}_{\varrho_N}^\nu(\Psi_0^{-\infty}(X; {}^0\Omega^{1/2})) \subseteq \mathfrak{N}$.

On the other hand, let $(\mathcal{N}, \mathcal{F}) \in \mathfrak{N}$ be arbitrary. By a partition of unity, we can assume that $\mathcal{F}$ is supported in $T^*\partial X|_{V^{\partial X}} \cong V^{\partial X} \times \mathbb{R}_\eta^{n-1}$ for some local coordinates $\chi : \partial X \supseteq V^{\partial X} \to \mathbb{R}_y^{n-1}$ on the boundary. Choose a cut-off function $\omega \in \mathcal{C}_c^\infty(\overline{\mathbb{R}}_+)$ with $\omega(r) = 1$ near $r = 0$, and define

$$\widehat{\kappa}(\tau, U, r, y') := \omega(r) \int_{\mathbb{R}_\eta^{n-1}} e^{iU\eta} \mathcal{F}(y', \eta; \tau) d\eta.$$

Then we have $\widehat{\kappa} \in \dot{\mathcal{C}}^\infty([-1,1]) \widehat{\otimes}_\pi \mathscr{S}(\mathbb{R}_U^{n-1}) \widehat{\otimes}_\pi \mathcal{C}_c^\infty(\overline{\mathbb{R}}_+) \widehat{\otimes}_\pi \mathcal{C}_c^\infty(\mathbb{R}_{y'}^{n-1})$, and we easily see that $\kappa = \widehat{\kappa}(\tau, U, r, y') \left| d\tau \, dU \frac{dr}{r^n} dy' \right|^{\frac{1}{2}}$ is the lifted Schwartz kernel of some 0-operator $A \in \Psi_0^{-\infty}(X; {}^0\Omega^{1/2})$ with $\mathcal{F}_{\varrho_N}^\nu(A) = \mathcal{F}$ and $\mathcal{N}_{\varrho_N}^\nu(A) = \mathcal{N}$. $\square$

The case $\Psi_0^m(X; {}^0\Omega^{1/2})$.

Let us begin with some preparatory lemmata about classical symbols that can be checked by direct computation. Throughout this part, $\chi \in \mathcal{C}_c^\infty(\overline{\mathbb{R}}_+)$ satisfies $\chi(r) = 1$ near $r = 0$.

LEMMA 4.4.2. *Let $a \in S_{cl}^m(\overline{\mathbb{R}}_+ \times \mathbb{R}_y^{n-1}; \mathbb{R}_\xi \times \mathbb{R}_\eta^{n-1})$ be arbitrary. Then the map*

$$\mathbb{R}_y^{n-1} \times \mathbb{R}_\eta^{n-1} \ni (y, \eta) \longmapsto [(r, \xi) \longmapsto \chi(r) a(0, y, \xi, r\eta)] \in S_{cl}^m(\overline{\mathbb{R}}_+; \mathbb{R}_\xi)$$

is well-defined and smooth with respect to the canonical Fréchet topology on the space $S_{cl}^m(\overline{\mathbb{R}}_+; \mathbb{R}_\xi)$.

Let $C_0 > 1$, $R > 0$, $1/R < \tau_1 < 1$, and $\omega \in \mathcal{C}_c^\infty(\overline{\mathbb{R}}_+ \times \mathbb{R}_T)$ be a smooth function with $\omega(s, T) = 1$ for $(s, T) \in \{|T| \leq C_0, 0 \leq s \leq C_0, \text{ and } |sT| \leq 1/R\}$, and $\operatorname{supp} \omega \subseteq \{|sT| \leq \tau_1\}$.

LEMMA 4.4.3. *Let $a \in S_{cl}^m(\overline{\mathbb{R}}_+ \times \mathbb{R}_y^{n-1}; \mathbb{R}_\xi \times \mathbb{R}_\eta^{n-1})$ be arbitrary. Then for each $(y, \eta) \in \mathbb{R}_y^{n-1} \times (\mathbb{R}_\eta^{n-1} \setminus \{0\})$ the map*

$$b(y, \eta) : \mathbb{R}_+ \times \mathbb{R}_T \times \mathbb{R}_\xi \ni (s, T, \xi) \longmapsto s^m \omega(s, T) a(0, y, \xi/s, \eta/s)$$

extends uniquely to a classical symbol $b(y, \eta) \in S_{cl}^m(\overline{\mathbb{R}}_+ \times \mathbb{R}_T; \mathbb{R}_\xi)$, and the map $b : \mathbb{R}_y^{n-1} \times (\mathbb{R}_\eta^{n-1} \setminus \{0\}) \to S_{cl}^m(\overline{\mathbb{R}}_+ \times \mathbb{R}_T; \mathbb{R}_\xi)$ is smooth with respect to the canonical Fréchet-topology on $S_{cl}^m(\overline{\mathbb{R}}_+ \times \mathbb{R}_T; \mathbb{R}_\xi)$. Moreover, we have
(4.15)
$$\sigma^{(m)}(b(y,\eta))(s, T, \xi) = \omega(s, T) \sigma^{(m)}(a)(0, y, \xi, 0) \text{ for } (s, T, \xi) \in \overline{\mathbb{R}}_+ \times \mathbb{R}_T \times (\mathbb{R}_\xi \setminus \{0\}),,$$

where $\sigma^{(m)}$ denotes the homogeneous principal part of a classical symbol.

Proof: Let $a \sim \sum_{j=0}^\infty a_{m-j}$ with $a_{m-j} \in S^{[m-j]}(\overline{\mathbb{R}}_+ \times \mathbb{R}_y^{n-1}; \mathbb{R}_\xi \times \mathbb{R}_\eta^{n-1})$ be the homogeneous asymptotic expansion of a. Then we get for all small $s > 0$ and all $N \in \mathbb{N}_0$ with $N > m$
(4.16)
$$b(y, \eta)(s, T, \xi) = \omega(s, T) \left(\sum_{j=0}^N s^j a_{m-j}(0, y, \xi, \eta) + s^m R_{m-N-1}(0, y, \xi/s, \eta/s) \right)$$

with $R_{m-N-1} \in S^{m-N-1}(\overline{\mathbb{R}}_+ \times \mathbb{R}_y^{n-1}; \mathbb{R}_\xi \times \mathbb{R}_\eta^{n-1})$. Since the remainder in (4.16) extends to an element in $\dot{\mathcal{C}}^N(\overline{\mathbb{R}}_+ \times \mathbb{R}_T \times \mathbb{R}_\xi)$, the space of N-times differentiable functions on $\overline{\mathbb{R}}_+ \times \mathbb{R}_T \times \mathbb{R}_\xi$ vanishing up to order N at $\{s = 0\}$, for all $N \in \mathbb{N}$, we obtain $b(y, \eta) \in \mathcal{C}^\infty(\overline{\mathbb{R}}_+ \times \mathbb{R}_T \times \mathbb{R}_\xi)$.

A Taylor expansion of a_{m-j} with respect to η together with a version of [**36**, Proposition 18.1.4] then even gives $b(y,\eta) \in S_{cl}^m(\overline{\mathbb{R}}_+ \times \mathbb{R}_T; \mathbb{R}_\xi)$; the homogeneous components of $b(y,\eta)$ are given by $b_{m-k}(y,\eta) : \overline{\mathbb{R}}_+ \times \mathbb{R}_T \times (\mathbb{R}_\xi \setminus \{0\}) \to \mathbb{C}$ with

$$(4.17) \quad b_{m-k}(y,\eta) : (s,T,\xi) \longmapsto \omega(s,T) \sum_{j=0}^{k} s^j \sum_{|\alpha|=k-j} \eta^\alpha (\partial_\eta^\alpha a_{m-j})(0,y,\xi,0).$$

In particular, for $k=0$, we obtain (4.15).

The homogeneous parts in (4.17) obviously depend smoothly on (y,η), and a careful inspection of the corresponding remainders shows that they depend smoothly on (y,η) as well. This completes the proof. □

Using the isomorphism (4.10), and the identifications (1.2) and (1.3) given by the boundary defining function ϱ_N, we, thus, obtain an identification

$$(4.18) \quad j_{\varrho_N} : T^*\partial X \oplus \partial X \times \mathbb{R} \longrightarrow {}^0T^*\partial X \oplus \left({}^0TX|_{\partial X}/{}^0T\partial X\right)^* \cong {}^0T^*X|_{\partial X}$$

as vector bundles over ∂X.

PROPOSITION 4.4.4. *For $\eta \in T^*\partial X \setminus \{0\}$ and $A \in \Psi_0^m(X; {}^0\Omega^{1/2})$, the operator $\mathcal{N}_{\varrho_N}^\nu(A)(\eta)$ extends uniquely to an element in $\Psi_{b,c}^{m,m}(M; {}^{b,c}\Omega^{1/2})$, and the map*

$$\mathcal{N}_{\varrho_N}^\nu(A) : T^*\partial X \setminus \{0\} \longrightarrow \Psi_{b,c}^{m,m}(M; {}^{b,c}\Omega^{1/2}))$$

is smooth with uniform estimates whenever η is restricted to a compact subset of $T^*\partial X \setminus \{0\}$. Moreover, we have for all $\eta \in T^*\partial X \setminus \{0\}$, all $(z,\zeta) \in {}^{b,c}T^*M \setminus \{0\}$, all $\xi \in \mathbb{R}$, and all $w \in \mathbb{C}$

$$(4.19) \quad {}^{b,c}\sigma^{(m,m)}(\mathcal{N}_{\varrho_N}^\nu(A)(\eta))(z,\zeta) = {}^0\sigma^{(m)}(A)\left(j_{\varrho_N}(0,(\pi(\eta),\zeta))\right),$$

$$(4.20) \quad I_c^{(m)}(\mathcal{N}_{\varrho_N}^\nu(A)(\eta))(\xi) = {}^0\sigma^{(m)}(A)\left(j_{\varrho_N}(\eta,(\pi(\eta),\xi))\right), \text{ and}$$

$$(4.21) \quad I_b\left(\mathcal{N}_{\varrho_N}^\nu(A)(\eta)\right)(w) = I_A(w - i\frac{n-1}{2})(\pi(\eta)).$$

Proof: Let us first consider the blow-down map $\beta : M_{b,c}^2 \xrightarrow{\varphi_{b,c}} \widetilde{M}_{b,c}^2 \xrightarrow{\beta} M_N$ where we have used the notations from Proposition 3.2.1. It is straight-forward to check that the lifted Schwartz kernels of operators in $\Psi_{b,c}^{m,m}(M; {}^{b,c}\Omega^{1/2})$ correspond under pushing-forward with β_* exactly to the following subspace
$$(4.22)$$
$$\varrho_{ff^{M_N}}^{-m} \left\{ \kappa \in I_{cl}^m\left(M_N, \Delta_{M_N}; \varrho_{ff^b}^{-1/2} \varrho_{ff^{M_N}}^{-1} \Omega^{1/2}(M_N)\right) : \kappa \equiv \partial M_N \setminus (\text{ff}^b \cup \text{ff}^{M_N}) \right\}$$

of $\mathcal{C}^{-\infty}(M_N, \varrho_{ff^b}^{-1/2} \varrho_{ff^{M_N}}^{-1} \Omega^{1/2}(M_N))$, where ff^b stands for the face of M_N coming from $\{r=0\}$. For brevity, we write I_{M_N} for the space in (4.22).

Let $A \in \Psi_0^m(X; {}^0\Omega^{1/2})$ be arbitrary. By Proposition 4.4.1 we can assume that the lifted Schwartz kernel κ_A of the 0-operator A is supported in a small neighborhood of $\Delta_{0,e} \cap \text{ff}^{0,e}$; by a partition of unity we can further assume that $\kappa_A = \widehat{\kappa}_A(\tau, U, r, y') |d\tau\, dU \frac{dr}{r^n} dy'|^{\frac{1}{2}}$ with

$$\operatorname{supp} \widehat{\kappa} \subseteq (-1/R, 1/R) \times \{|U| \leq R\} \times [0, R) \times \{|y'| \leq R\}$$

for some $R > 0$, and

$$\widehat{\kappa}_A(\tau, U, r, y') = {}^{Os}\!\!-\!\!\int_{\mathbb{R}_\eta^{n-1}} \int_{\mathbb{R}_\xi} e^{iU\eta} e^{i\tau\xi} a(r, y', \xi, \eta) d\xi d\eta$$

4.4. CHARACTERIZATION OF THE REDUCED NORMAL OPERATOR

for a symbol $a \in S_{cl}^m(\overline{\mathbb{R}}_+ \times \mathbb{R}_{y'}^{n-1}; \mathbb{R}_\xi \times \mathbb{R}_\eta^{n-1})$; thus, we obtain for the reduced normal operator

$$(4.23) \qquad \mathcal{N}_{\varrho_N}^\nu(A)(y,\eta;\tau,r) = {}^{Os\text{-}}\!\!\int_{\mathbb{R}_\xi} e^{i\tau\xi} a(0,y,\xi,r\eta) d\xi \left|\frac{dr}{r}d\tau\right|^{\frac{1}{2}}$$

with $\operatorname{supp} \mathcal{N}_{\varrho_N}^\nu(A)(y,\eta) \subseteq (-1/R, 1/R) \times \overline{\mathbb{R}}_+$ uniformly in y and η.

The kernel in (4.23) has the correct conormal behavior (4.22) for $r > 0$ by the very definition, and near ff^b by Lemma 4.4.2; as in the proof of Proposition 4.4.1, (4.21) is just a combination of (2.15), (3.6), and (4.4). It remains to consider $r \to \infty$, i.e. the behavior near ff^{M_N}. Introducing for large $r > 0$ the coordinate $s = 1/r$ identifies the reduced normal operator with the distributional density

$$\mathcal{N}_{\varrho_N}^\nu(A)(y,\eta) = {}^{Os\text{-}}\!\!\int_{\mathbb{R}_\xi} e^{i\tau\xi} a(0,y,\xi,\eta/s) d\xi \left|\frac{ds}{s}d\tau\right|^{\frac{1}{2}},$$

and $\{s=0\}$ with the face $[-1,1] \times \{1\}$ of $[-1,1] \times [0,1]$. Therefore, $\mathcal{N}_{\varrho_N}^\nu(A)$ has the correct behavior (4.22) near the faces lb and rb of M_N, i.e. for $|\tau| \geq 1/R$.

Let $\beta : M_N \to [-1,1] \times [0,1]$ be the blow-down map, and denote the lift of $\mathcal{N}_{\varrho_N}^\nu(A)$ under β again by $\mathcal{N}_{\varrho_N}^\nu(A)$. By a partition of unity on M_N it suffices to consider the cases that $\mathcal{N}_{\varrho_N}^\nu(A)(y,\eta)$ is supported uniformly in (y,η)

(i) near $\mathrm{ff}^{M_N} \cap \Delta_{M_N}$,

(ii) near $F_{M_N,+} \cap \mathrm{ff}^{M_N}$, and

(iii) near $F_{M_N,-} \cap \mathrm{ff}^{M_N}$.

In case (i), we can use the coordinates $s, T := \tau/s$ near $\mathrm{ff}^{M_N} \cap \Delta_{M_N}$, thus, we have $\mathrm{ff}^{M_N} = \{s=0\}$, $\Delta_{M_N} = \{T=0\}$, and the lift of $\mathcal{N}_{\varrho_N}^\nu(A)$ corresponds to the distributional density

$$\mathcal{N}_{\varrho_N}^\nu(A)(y,\eta) = {}^{Os\text{-}}\!\!\int_{\mathbb{R}_\xi} e^{isT\xi} a(0,y,\xi,\eta/s) d\xi \, s \left|\frac{ds}{s^2}d\tau\right|^{\frac{1}{2}}.$$

By (i), there exists $C_0 > 0$ such that

$$\operatorname{supp} \mathcal{N}_{\varrho_N}^\nu(A)(y,\eta) \subseteq \{|T| \leq C_0, 0 \leq s \leq C_0, \text{ and } |sT| \leq 1/R\}$$

for all $(y,\eta) \in \mathbb{R}_y^{n-1} \times \mathbb{R}_\eta^{n-1} \setminus \{0\}$. Thus, we get

$$\mathcal{N}_{\varrho_N}^\nu(A)(y,\eta) = {}^{Os\text{-}}\!\!\int_{\mathbb{R}_\xi} e^{iT\xi} a(0,y,\xi/s,\eta/s) d\xi \left|\frac{ds}{s^2}d\tau\right|^{\frac{1}{2}},$$

and Lemma 4.4.3 gives (4.19) and (4.22) in case (i). For (4.20), it suffices to note that $\left(\varrho_{\mathrm{ff}^{M_N}}^m \mathcal{N}_{\varrho_N}^\nu(A)(y,\eta)\right)|_{\mathrm{ff}^{M_N}}$ is given by the distributional density

$${}^{Os\text{-}}\!\!\int_{\mathbb{R}_\xi} e^{iT\xi} \sigma^{(m)}(a)(0,y,\xi,\eta) d\xi \, |dT|^{\frac{1}{2}},$$

hence, $I_c^{(m)}\left(\mathcal{N}_{\varrho_N}^\nu(A)(y,\eta)\right)(\xi) = \sigma^{(m)}(a)(0,y,\xi,\eta)$ by (3.7).

Near the corner $F_{M_N,+} \cap \mathrm{ff}^{M_N}$ of codimension 2 we use the coordinates $S := s/\tau$ and $\tau \geq 0$; by (ii) we can assume $s = S\tau \leq 1$, and we have $F_{M_N,+} = \{S = 0\}$ and

$\mathrm{ff}^{M_N} = \{\tau = 0\}$; the lift of $\mathcal{N}^\nu_{\varrho_N}(A)(y,\eta)$ then has the form

$$\mathcal{N}^\nu_{\varrho_N}(A)(y,\eta) = {}^{Os\text{-}}\!\!\int_{\mathbb{R}_\xi} e^{i\tau\xi} a(0,y,\xi,\frac{\eta}{S\tau})d\xi \, \frac{\tau}{S^{1/2}} \left| dS \frac{d\tau}{\tau^2} \right|^{\frac{1}{2}}.$$

Let $a \sim \sum_{j=0}^\infty a_{m-j}$ with $a_{m-j} \in S^{[m-j]}(\overline{\mathbb{R}}_+ \times \mathbb{R}^{n-1}_y; \mathbb{R}_\xi \times \mathbb{R}^{n-1}_\eta)$ be the homogeneous asymptotic expansion of a. Then we get for any $L \in \mathbb{N}_0$

$$\tau^m S^{-L} \, {}^{Os\text{-}}\!\!\int_{\mathbb{R}_\xi} e^{i\tau\xi} a(0,y,\xi,\frac{\eta}{S\tau})d\xi \, \frac{\tau}{S^{1/2}} =$$

$$= \sum_{j=0}^N \tau^j S^{j-m-L-3/2} \underbrace{{}^{Os\text{-}}\!\!\int_{\mathbb{R}_\xi} e^{i\xi/S} a_{m-j}(0,y,\xi,\eta)d\xi}_{:=H_j(y,\eta;\tau,S)} +$$

$$+ \tau^m S^{-L-3/2} \underbrace{{}^{Os\text{-}}\!\!\int_{\mathbb{R}_\xi} e^{i\xi/S} R_{m-N-1}(0,y,\frac{\xi}{S\tau},\frac{\eta}{S\tau})d\xi}_{=:\widetilde{H}_{N+1}(y,\eta;\tau,S)},$$

where $N \in \mathbb{N}_0$ is arbitrary with $N > m$ and the remainder R_{m-N-1} satisfy $R_{m-N-1} \in S^{m-N-1}(\overline{\mathbb{R}}_+ \times \mathbb{R}^{n-1}_y; \mathbb{R}_\xi \times \mathbb{R}^{n-1}_\eta)$.

Using the identity $i/S e^{i\xi/S} = \partial_\xi e^{i\xi/S}$ to regularize the oscillatory integral, we get for each $M \in \mathbb{N}$ with $M > m + 1 - j$

$$H_j(y,\eta;\tau,S) = i^M \tau^j S^{M+j-m-L-3/2} \int_{\mathbb{R}_\xi} e^{i\xi/S} \left(\partial_\xi^M a_{m-j} \right)(0,y,\xi,\eta) d\xi;$$

since we can take M arbitrary large, $H_j(y,\eta) \in \mathcal{C}^\infty([0,1) \times \overline{\mathbb{R}}_+)$ vanishes with all derivatives at $S = 0$.

On the other hand, because of

$$\text{const}\,(1+|\xi|^2) \leq \tau^2 S^2 + |\xi|^2 + |\eta|^2 \leq \text{const}\,'(1+|\xi|^2),$$

we obtain $\widetilde{H}_{N+1}(y,\eta) \in \mathcal{C}^k([0,1) \times \overline{\mathbb{R}}_+)$ provided $N \geq L + m + 1 + 2k$. Since N can be chosen arbitrary large, $\mathcal{N}^\nu_{\varrho_N}(A)(y,\eta)$ satisfies (4.22), i.e. $\mathcal{N}^\nu_{\varrho_N}(A)(y,\eta)$ is smooth up to ff^{M_N} and vanishes to infinite order at $F_{M_N,+}$. Moreover, a careful inspection of the functions H_j and $\widetilde{H}_{N+1}$ shows that they depend smoothly on $(y,\eta) \in \mathbb{R}^{n-1}_y \times (\mathbb{R}^{n-1}_\eta \setminus \{0\})$ as well.

The case (iii) follows by symmetry from (ii). This completes the proof. □

As a simple illustration, let us compute the reduced normal operator of a multiplication operator.

EXAMPLE 4.4.5. Let $\varphi \in \mathcal{C}^\infty(X)$ be arbitrary and $M_\varphi \in \Psi^0_0(X; {}^0\Omega^{1/2})$ be the operator of multiplication with φ. Then we have for all $\eta \in T^*\partial X \setminus \{0\}$

$$\mathcal{N}^\nu_{\varrho_N}(M_\varphi)(\eta) = \varphi(\pi(\eta)) \mathrm{id}_{\dot{\mathcal{C}}^\infty(M,{}^{b,c}\Omega^{1/2})}.$$

The case $\Psi_0^{-\infty,0,\gamma}(X; {}^0\Omega^{1/2})$.

For a convenient description of the properties of $\mathcal{F}^\nu_{\varrho_N}(A)(\eta)$ as $|\eta| \to 0, \infty$, we need two auxiliary spaces. Let $\overline{T}^*\partial X$ be the radial compactification of $T^*\partial X$, and $\beta : \mathcal{P}^*\partial X := [\overline{T}^*\partial X; \partial X] \to \overline{T}^*\partial X$ be the blow up of the zero section $\partial X \hookrightarrow \overline{T}^*\partial X$ in $\overline{T}^*\partial X$. Note that the choice of a homogeneous function $T^*\partial X \setminus \{0\} \to \mathbb{R}_+$ of degree one together with (3.1) yields a diffeomorphism $\mathcal{P}^*\partial X \to S^*\partial X \times [0,1]$

4.4. CHARACTERIZATION OF THE REDUCED NORMAL OPERATOR

where $S^*\partial X := (T^*\partial X \setminus \{0\})/\mathbb{R}_+$ is the *cosphere bundle* of the boundary ∂X. Let $\varrho_0 : \mathcal{P}^*\partial X \to \overline{\mathbb{R}}_+$ resp. $\varrho_\infty : \mathcal{P}^*\partial X \to \overline{\mathbb{R}}_+$ be a defining function for the face that corresponds to $S^*\partial X \times \{0\}$ resp. $S^*\partial X \times \{\infty\}$ under this identification. Note that $S^*\partial X$ has two components if the boundary ∂X is one-dimensional. We continue to denote the canonical projection $\mathcal{P}^*\partial X \to \partial X$ with π.

Let us start with some basic observations. For simplicity, we restrict ourselves to the case $\gamma_F > n - 1$.

PROPOSITION 4.4.6. *Let $\gamma = (\gamma_T, \gamma_B, \gamma_F)$ be a weight system with $\gamma_F > n - 1$, and $A \in \Psi_0^{-\infty,0,\gamma}(X; {}^0\Omega^{1/2})$. Then we have $\mathcal{F}_{\varrho_N}^\nu(A)(\eta) \in \mathcal{A}^{(\gamma_T,\gamma_B)}([-1,1])$ for any $\eta \in T^*\partial X$, and, if $\eta \neq 0 = \pi(\eta)$, $\mathcal{N}_{\varrho_N}^\nu(A)(\eta)$ extends uniquely to an element in $\Psi_{b,c}^{-\infty,-\infty,(\gamma_T,\gamma_B)}(M; {}^{b,c}\Omega^{1/2})$ such that*

$$(4.24) \quad I_c^{(k)}(\mathcal{N}_{\varrho_N}^\nu(A)(\eta)) = 0 \text{ for all } k \in \mathbb{Z}, \text{ and}$$

$$(4.25) \quad I_b\left(\mathcal{N}_{\varrho_N}^\nu(A)(\eta)\right)(z) = \int_{-1}^{1} \left(\frac{1+\tau}{1-\tau}\right)^{-iz} \mathcal{F}_{\varrho_N}^\nu(A)(\pi(\eta); \tau) \frac{d\tau}{(1-\tau^2)^{1/2}}$$

for all $z \in \mathbb{C}$ with $-1/2 - \gamma_T < \text{Im } z < 1/2 + \gamma_B$. Moreover, we have

$$\mathcal{N}_{\varrho_N}^\nu(A) \in \mathcal{C}^\infty\left(T^*\partial X \setminus \{0\}, \Psi_{b,c}^{-\infty,-\infty,(\gamma_T,\gamma_B)}(M; {}^{b,c}\Omega^{1/2})\right).$$

The map $\mathcal{F}_{\varrho_N}^\nu(A) : T^*\partial X \to \mathcal{A}^{(\gamma_T,\gamma_B)}([-1,1])$ is continuous, and $\beta^* \mathcal{F}_{\varrho_N}^\nu(A)|_{\{\varrho_\infty > 0\}}$ extends to a continuous map $\beta^* \mathcal{F}_{\varrho_N}^\nu(A) : \mathcal{P}^*\partial X \to \mathcal{A}^{(\gamma_T,\gamma_B)}([-1,1])$ that is smooth on $\{\varrho_0 > 0\}$ and vanishes with all derivatives at the boundary face(s) corresponding to $\{\varrho_\infty = 0\}$.

Proof: Again, we are using the local formulas (4.4) and (4.5) for $\mathcal{F}_{\varrho_N}^\nu(A)(\eta)$ and $\mathcal{N}_{\varrho_N}^\nu(A)(\eta)$. By (2.14), $\widehat{\kappa}_A$ is symbolic of finite order with respect to U, thus, the Fourier transform $\mathcal{F}_{\varrho_N}^\nu(A)(\eta; \tau)$ with respect to U is smooth outside the zero section ($\eta = 0$), and rapidly decreasing for $|\eta| \to \infty$. In particular, for $\eta \in T^*\partial X \setminus \{0\}$ fixed, the kernel $\mathcal{N}_{\varrho_N}^\nu(A)(\eta)(\tau, r) = \mathcal{F}_{\varrho_N}^\nu(A)(r\eta; \tau)$ is rapidly decreasing as $r \to \infty$. This gives $\mathcal{N}_{\varrho_N}^\nu(A)(\eta) \in \Psi_{b,c}^{-\infty,-\infty,(\gamma_T,\gamma_B)}(M; {}^{b,c}\Omega^{1/2})$ and (4.24). For (4.25), it suffices to combine (3.6), and (4.4).

Because of $\gamma_F > n - 1$ the integral in (4.5) is absolutely convergent by (2.14), hence $\mathcal{F}_{\varrho_N}^\nu(A)$ is continuous on $T^*\partial X$. $\square$

For an almost characterization of the range of the reduced normal operator on $\Psi_0^{-\infty,0,\gamma}(X; {}^0\Omega^{1/2})$ we need the following spaces.

DEFINITION 4.4.7. For $\gamma_T, \gamma_B \in \mathbb{R}$ and $0 < \delta \leq 1$, let $\widetilde{\mathfrak{N}}_0^{-\infty,0,(\gamma_T,\gamma_B)}(\delta)$ be the space of all pairs $(\mathcal{N}, \mathcal{F})$ of smooth functions $\mathcal{F} : T^*\partial X \setminus \{0\} \to \mathcal{A}^{(\gamma_T,\gamma_B)}([-1,1])$ and $\mathcal{N} : T^*\partial X \setminus \{0\} \to \Psi_{b,c}^{-\infty,-\infty,(\gamma_T,\gamma_B)}(M; {}^{b,c}\Omega^{1/2})$ satisfying

(a) $\mathcal{N}(\eta; \tau, r) = \mathcal{F}(r\eta; \tau) \left|d\tau \frac{dr}{r}\right|^{\frac{1}{2}}$ for all $\eta \in T^*\partial X$, all $\tau \in (-1, 1)$, and all $r \in \mathbb{R}_+$.

(b) The map $\mathcal{F}$ extends to a continuous map $\mathcal{F} : T^*\partial X \to \mathcal{A}^{(\gamma_T,\gamma_B)}([-1,1])$ satisfying

$$(4.26) \quad I_b\left(\mathcal{N}(\eta)\right)(z) = \int_{-1}^{1} \left(\frac{1+\tau}{1-\tau}\right)^{-iz} \mathcal{F}(\pi(\eta); \tau) \frac{d\tau}{(1-\tau^2)^{1/2}}$$

for all $z \in \mathbb{C}$ with $-1/2 - \gamma_T < \text{Im } z < 1/2 + \gamma_B$.

(c) The map $\beta^*\mathcal{F}|_{\{\varrho_0>0\}} : \{\varrho_0 > 0\} \to \mathcal{A}^{(\gamma_T,\gamma_B)}([-1,1])$ is smooth, and vanishes with all derivatives at the boundary faces of $\mathcal{P}^*\partial X$ corresponding to $\{\varrho_\infty = 0\}$.

(d) For each $P \in \mathrm{Diff}_b^*(\mathcal{P}^*\partial X)$, the set

(4.27) $$\{\varrho_0^{-\delta} P\left(\beta^*\mathcal{F}(\vartheta) - \mathcal{F}(\pi(\vartheta))\right) : \vartheta \in \mathcal{P}^*\partial X\} \subseteq \mathcal{A}^{(\gamma_T,\gamma_B)}([-1,1])$$

is bounded.

The spaces $\widetilde{\mathfrak{N}}_0^{-\infty,0,(\gamma_T,\gamma_B)}(\delta)$ clearly do not depend on the particular choice of the defining functions ϱ_0 and ϱ_∞.

Let $\mathfrak{N}_0^{-\infty,0,\gamma} := \{(\mathcal{N}_{\varrho_N}^\nu(A), \mathcal{F}_{\varrho_N}^\nu(A)) : A \in \Psi_0^{-\infty,0,\gamma}(X; {}^0\Omega^{1/2})\}$ be the range of the reduced normal operator and the function $\mathcal{F}_{\varrho_N}^\nu$ on $\Psi_0^{-\infty,0,\gamma}(X; {}^0\Omega^{1/2})$; the following theorem is more or less a parameter-dependent version of the results of Appendix A, and we use the notations introduced there.

THEOREM 4.4.8. *Let $\gamma = (\gamma_T, \gamma_B, \gamma_F)$ be an arbitrary weight system for X with $n - 1 < \gamma_F < n$. Then we have*
(4.28)
$$\bigcup_{\gamma_F-(n-1)<\delta\leq 1} \widetilde{\mathfrak{N}}_0^{-\infty,0,(\gamma_T,\gamma_B)}(\delta) \subseteq \mathfrak{N}_0^{-\infty,0,\gamma} \subseteq \bigcap_{0<\delta<\gamma_F-(n-1)} \widetilde{\mathfrak{N}}_0^{-\infty,0,(\gamma_T,\gamma_B)}(\delta).$$

Proof: Because of (4.4) and (a) in Definition 4.4.7 it suffices to consider $\mathcal{F}$ and $\mathcal{F}_{\varrho_N}^\nu(A)$.

Using local coordinates on ∂X, a partition of unity, and a homogeneous function of degree one on $T^*\partial X \setminus \{0\}$, we can assume that the blow down map β is of the form

(4.29) $$\beta| : \mathcal{P}^*\partial X \setminus \{\varrho_\infty = 0\} = \mathbb{R}_y^{n-1} \times S^{n-2} \times \overline{\mathbb{R}}_+ \longrightarrow \mathbb{R}_y^{n-1} \times \mathbb{R}_\eta^{n-1} = T^*\partial X :$$
$$(y, \hat{\eta}, \varrho) \longmapsto (y, \varrho\hat{\eta}),$$

and that $\mathcal{F}$ satisfies $\mathcal{F}(y, \eta; \tau) = 0$ whenever $|y| \geq R$ for some large R.

First, let $\gamma_F - (n-1) < \delta \leq 1$ and $(\mathcal{N}, \mathcal{F}) \in \widetilde{\mathfrak{N}}_0^{-\infty,0,(\gamma_T,\gamma_B)}(\delta)$ be arbitrary. Choose $\omega \in \mathcal{C}^\infty(\overline{\mathbb{R}}_+)$ with $\omega = 1$ near $r = 0$. We are going to show that the density κ with $\kappa = \hat{\kappa}(\tau, U, r, y') \left| d\tau\, dU\, \frac{dr}{r^n} dy' \right|^{\frac{1}{2}}$ where

(4.30) $$\hat{\kappa}(\tau, U, r, y') := \omega(r) \int_{\mathbb{R}_\eta^{n-1}} e^{iU\eta} \mathcal{F}(y', \eta; \tau) d\eta$$

is the lifted Schwartz kernel of an operator $A \in \Psi_0^{-\infty,0,\gamma}(X; {}^0\Omega^{1/2})$ satisfying $\mathcal{F}_{\varrho_N}^\nu(A) = \mathcal{F}$. By (4.5) and (4.30) it is sufficient to check $A \in \Psi_0^{-\infty,0,\gamma}(X; {}^0\Omega^{1/2})$.

For $\chi \in \mathcal{C}_c^\infty(\overline{\mathbb{R}}_+)$ with $\chi = 1$ near $r = 0$ and $\chi = 0$ for $r \geq R$, consider the following decomposition

$$\hat{\kappa}(\tau, U, r, y') =$$
$$\underbrace{\omega(r) \int_{\mathbb{R}_\eta^{n-1}} e^{iU\eta}(1 - \chi(|\eta|))\mathcal{F}(y', \eta; \tau) d\eta}_{=:\hat{\kappa}_1(\tau,U,r,y')} + \underbrace{\omega(r) \mathcal{F}(y', 0; \tau) \int_{\mathbb{R}_\eta^{n-1}} e^{iU\eta} \chi(|\eta|) d\eta}_{=:\hat{\kappa}_2(\tau,U,r,y')}$$
$$+ \underbrace{\omega(r) \int_{\mathbb{R}_\eta^{n-1}} e^{iU\eta} \chi(|\eta|) \left(\mathcal{F}(y', \eta; \tau) - \mathcal{F}(y', 0; \tau)\right) d\eta}_{=:\hat{\kappa}_3(\tau,U,r,y')}$$

4.4. CHARACTERIZATION OF THE REDUCED NORMAL OPERATOR

Since $\mathbb{R}_y^{n-1} \times \mathbb{R}_\eta^{n-1} \to \mathcal{A}^{(\gamma_T,\gamma_B)}([-1,1]) : (y,\eta) \mapsto (1-\chi(|\eta|))\mathcal{F}(y,\eta;\cdot)$ is rapidly decreasing with respect to η and compactly supported with respect to y, the kernel κ_1 satisfies (2.14) with $\gamma_F = \infty$, i.e. κ_1 corresponds to an operator in the space $\Psi_0^{-\infty,0,(\gamma_T,\gamma_B,\infty)}(X; {}^0\Omega^{1/2})$.

The kernel κ_2 corresponds to an operator in $\Psi_0^{-\infty,0,(\gamma_T,\gamma_B,\infty)}(X; {}^0\Omega^{1/2})$ because of $[y' \mapsto \mathcal{F}(y',0;\cdot)] \in \mathcal{C}_c^\infty(\mathbb{R}_{y'}^{n-1}, \mathcal{A}^{(\gamma_T,\gamma_B)}([-1,1]))$ and $[\eta \mapsto \chi(|\eta|)] \in \mathcal{C}_c^\infty(\mathbb{R}_\eta^{n-1})$.

Next, we are going to show that for all $k, \ell \in \mathbb{N}_0$ and all multi-indices $\beta \in \mathbb{N}_0^{n-1}$, the inverse Fourier-transform $\mathcal{F}_{\eta \to U}^{-1}$ of

$$g : (y, \eta, r, \tau) \longmapsto (1+\tau)^{-\gamma_T}(1-\tau)^{-\gamma_B}$$
$$[(1-\tau^2)\partial_\tau]^k \partial_r^\ell \partial_y^\beta \omega(r) \chi(|\eta|) \left(\mathcal{F}(y,\eta;\tau) - \mathcal{F}(y,0;\tau)\right)$$

with respect to η belongs to the space $S^{-\gamma_F}(;\mathbb{R}_U^{n-1})$ with uniform estimates in y, τ, and r; this implies $\kappa_3 \in \mathcal{K}_0^{-\infty,0,\gamma}(X_{0,e}^2, KD_{0,e}^{1/2})$, i.e. κ_3 corresponds to an element in $\Psi_0^{-\infty,0,\gamma}(X; {}^0\Omega^{1/2})$ as desired.

Let $\beta_p : \overline{\mathbb{R}}_+ \times S_{\widehat{\eta}}^{n-2} \longrightarrow \mathbb{R}_\eta^{n-1} : (\varrho, \widehat{\eta}) \longmapsto \eta = \varrho\widehat{\eta}$ be polar coordinates in the η-space. Recall that the map β_p corresponds exactly to blowing up the origin in $\mathbb{R}_\eta^{n-1} = T_y^* \partial X$. By (4.27), the map

$$G : \mathbb{R}_y^{n-1} \times (-1,1) \times \overline{\mathbb{R}}_+ \longrightarrow \varrho^\delta IL_R^\infty(\overline{\mathbb{R}}_+ \times S_{\widehat{\eta}}^{n-2}; \partial) :$$
$$(y, \tau, r) \longmapsto [(\varrho, \widehat{\eta}) \longmapsto g(y, \varrho\widehat{\eta}, \tau, r)]$$

is continuous and bounded in the sense that for each continuous seminorm q on the space $\varrho^\delta IL_R^\infty(\overline{\mathbb{R}}_+ \times S_{\widehat{\eta}}^{n-2}; \partial)$ there exists $C_q > 0$ with $q(G(y,\tau,r)) \leq C_q$ for all $(y, \tau, r) \in \mathbb{R}_y^{n-1} \times (-1,1) \times \overline{\mathbb{R}}_+$. Here $R > 0$ satisfies $\chi(\varrho) = 0$ for $\varrho \geq R$. Note that we have $G(y,\tau,r)(\varrho, \widehat{\eta}) = (\beta_p^* g(y,\cdot,\tau,r))(\varrho, \widehat{\eta})$. By Proposition A.13 the map

$$G_0 : \mathbb{R}_y^{n-1} \times (-1,1) \times \overline{\mathbb{R}}_+ \longrightarrow |\eta|^\delta IL_R^\infty(\mathbb{R}_\eta^{n-1}; \{0\}) : (y,\tau,r) \longmapsto [\eta \mapsto g(y,\eta,\tau,r)]$$

is then continuous and bounded in the same sense. Because of $\gamma_F - (n-1) < \delta \leq 1$ there exists $s \in \mathbb{R}$ with $0 < \gamma_F - \frac{n-1}{2} < s < \delta + \frac{n-1}{2}$. By Lemma A.9 we obtain the boundedness and continuity of

$$G_1 : \mathbb{R}_y^{n-1} \times (-1,1) \times \overline{\mathbb{R}}_+ \longrightarrow |\eta|^s IL_R^2(\mathbb{R}_\eta^{n-1}; \{0\}) : (y,\tau,r) \longmapsto [\eta \mapsto g(y,\eta,\tau,r)],$$

hence by Proposition A.11 the boundedness and continuity of

$$G_2 : \mathbb{R}_y^{n-1} \times (-1,1) \times \overline{\mathbb{R}}_+ \longrightarrow IH_R^s(\mathbb{R}_\eta^{n-1}; \{0\}) : (y,\tau,r) \longmapsto [\eta \mapsto g(y,\eta,\tau,r)],$$

and, finally, by Proposition A.6 the boundedness and continuity of

$$G_3 : \mathbb{R}_y^{n-1} \times (-1,1) \times \overline{\mathbb{R}}_+ \longrightarrow I_R^{-\gamma_F + \frac{n-1}{4}}(\mathbb{R}_\eta^{n-1}; \{0\}) : (y,\tau,r) \longmapsto [\eta \mapsto g(y,\eta,\tau,r)]$$

because of $-\gamma_F + \frac{n-1}{4} > -(s + \frac{n-1}{4})$. By the Definition A.3 of $I_R^{-\gamma_F + \frac{n-1}{4}}(\mathbb{R}_\eta^{n-1}; \{0\})$ this implies that the map

$$\mathbb{R}_y^{n-1} \times (-1,1) \times \overline{\mathbb{R}}_+ \longrightarrow S^{-\gamma_F}(;\mathbb{R}_U^{n-1}) : (y,\tau,r) \longmapsto \mathcal{F}_{\eta \to U}^{-1} g(y,\eta,\tau,r)$$

is well-defined, continuous and bounded, which completes the proof of the first inclusion in (4.28).

On the other hand, let $A \in \Psi_0^{-\infty,0,\gamma}(X; {}^0\Omega^{1/2})$ be arbitrary. By Proposition 4.4.6, it remains to show (4.27) in Definition 4.4.7. We use a partition of unity and local coordinates as in (4.29). By (2.14), the map

$$H_0 : (-1,1) \times \mathbb{R}_y^{n-1} \longrightarrow S^{-\gamma_F}(;\mathbb{R}_U^{n-1}) :$$
$$(\tau, y) \longmapsto \left[U \mapsto (1+\tau)^{-\gamma_T}(1-\tau)^{-\gamma_B}[(1-\tau^2)\partial_\tau]^k \partial_y^\beta \widehat{\kappa}_A(\tau, U, 0, y)\right]$$

is continuous and bounded for all $k \in \mathbb{N}_0$ and all multi-indices $\beta \in \mathbb{N}_0^{n-1}$. Choose $\chi \in \mathcal{C}_c^\infty(\overline{\mathbb{R}}_+)$ with $\chi = 1$ near $\varrho = 0$ and $\chi = 0$ for $\varrho \geq R$ for some $R > 0$. By Lemma A.4 the map

$$H_1 : (-1,1) \times \mathbb{R}_y^{n-1} \longrightarrow I_R^{-\gamma_F + \frac{n-1}{4}}(\mathbb{R}_\eta^{n-1}; \{0\}) :$$
$$(\tau, y) \longmapsto [\eta \mapsto \chi(|\eta|)\left(\mathcal{F}_{U \to \eta} H_0(\tau, y)\right)(\eta)]$$

is continuous and bounded, thus, the map

$$H_2 : (-1,1) \times \mathbb{R}_y^{n-1} \longrightarrow IH_R^s(\mathbb{R}_\eta^{n-1}; \{0\}) : (\tau, y) \longmapsto [\eta \mapsto H_1(\tau, y)(\eta)]$$

is continuous and bounded for all $s < \gamma_F - \frac{n-1}{2}$ by Proposition A.6. Using Proposition A.8 we see that for all $0 < \delta < \gamma_F - (n-1)$ the map

$$H_3 : (-1,1) \times \mathbb{R}_y^{n-1} \longrightarrow |\eta|^\delta IL_R^\infty(\mathbb{R}_\eta^{n-1}; \{0\}) :$$
$$(\tau, y) \longmapsto [\eta \mapsto \chi(|\eta|)(H_2(\tau, y)(\eta) - H_2(\tau, y)(0))]$$

is well-defined, continuous and bounded. Note that $H_2(\tau, y) \in IH_R^s(\mathbb{R}_\eta^{n-1}; \{0\})$ for all $s < \gamma_F - \frac{n-1}{2}$, in particular $H_2(\tau, y)$ is continuous in η. By Proposition A.13 the map

$$H_4 : (-1,1) \times \mathbb{R}_y^{n-1} \longrightarrow \varrho^\delta IL_R^\infty(\overline{\mathbb{R}}_+ \times S_{\widehat{\eta}}^{n-2}; \partial) : (\tau, y) \longmapsto \beta_p^* H_3(\tau, y)$$

is continuous and bounded for all $0 < \delta < \gamma_F - (n-1)$, where

$$\beta_p : \overline{\mathbb{R}}_+ \times S_{\widehat{\eta}}^{n-2} \longrightarrow \mathbb{R}_\eta^{n-1} : (\varrho, \widehat{\eta}) \longmapsto \varrho\widehat{\eta} = \eta$$

are polar coordinates in $\mathbb{R}_\eta^{n-1}$ as above. Note that we have

$$H_4(\tau, y)(\varrho, \widehat{\eta}) = \chi(\varrho)(1+\tau)^{-\gamma_T}(1-\tau)^{-\gamma_B}$$
$$\left[(1-\tau^2)\partial_\tau\right]^k \partial_y^\beta \left(\mathcal{F}_{\varrho_N}^\nu(A)(y, \varrho\widehat{\eta}; \tau) - \mathcal{F}_{\varrho_N}^\nu(A)(y, 0; \tau)\right).$$

Therefore, the second inclusion in (4.28) follows directly from the definition of $\varrho^\delta IL_R^\infty(\overline{\mathbb{R}}_+ \times S_{\widehat{\eta}}^{n-2}; \partial)$. □

4.5. Basic properties of the reduced normal operator

PROPOSITION 4.5.1. *Let $A_j \in \Psi_0^{m_j}(X; {}^0\Omega^{1/2})$, $j = 1, 2$, be arbitrary. Then we have for all $\eta \in T^*\partial X \setminus \{0\}$*

(4.31) $\quad \mathcal{N}_{\varrho_N}^\nu(A_1 A_2)(\eta) = \mathcal{N}_{\varrho_N}^\nu(A_1)(\eta)\mathcal{N}_{\varrho_N}^\nu(A_2)(\eta) \in \Psi_{b,c}^{m_1+m_2, m_1+m_2}(M; {}^{b,c}\Omega^{1/2})$.

Proof: Let $(x,y) : X \supseteq U \to \overline{\mathbb{R}}_+ \times \mathbb{R}_y^{n-1}$ be local coordinates near the boundary ∂X such that $\eta \in T^*\partial X$ corresponds to $(\overline{y}, \eta) \in \mathbb{R}_y^{n-1} \times \mathbb{R}_\eta^{n-1}$ in local coordinates. Furthermore, let $\kappa_j := \widehat{\kappa}_j(\tau, U, r, y') \left|d\tau\, dU\, \frac{dr}{r^n} dy'\right|^{\frac{1}{2}}$ be the local representation of

4.5. BASIC PROPERTIES OF THE REDUCED NORMAL OPERATOR

the lifted Schwartz kernel of $A_j \in \Psi_0^{m_j}(X; {}^0\Omega^{1/2})$. Then the lifted Schwartz kernel $\kappa = \widehat{\kappa}(\tau, U, r, y') \left| d\tau \, dU \frac{dr}{r^n} dy' \right|^{\frac{1}{2}}$ of $A_1 A_2$ satisfies

$$\widehat{\kappa}(\tau, U, 0, y') = \sqrt{2} \int_{\mathbb{R}^{n-1}_{U'}} \int_{-1}^{1} \frac{(1-\tau')^{n-1}}{(1-\tau\tau')^n} \widehat{\kappa}_1 \left(\frac{\tau - \tau'}{1 - \tau\tau'}, \frac{(1-\tau')U - (1-\tau)U'}{1 - \tau\tau'}, 0, y' \right) \widehat{\kappa}_2 (\tau', U', 0, y') \, d\tau' dU',$$

thus, we obtain for the reduced normal operator of $A_1 A_2$

$$\begin{aligned} \mathcal{N}^\nu_{\varrho_N}(A_1 A_2)(\bar{y}, \eta; \tau, r) &= \sqrt{2} \int_{-1}^{1} \frac{1}{1 - \tau\tau'} \mathcal{N}^\nu_{\varrho_N}(A_1) \left(\bar{y}, \eta; \frac{\tau - \tau'}{1 - \tau\tau'}, \frac{1 - \tau\tau'}{1 - \tau'} r \right) \\ &\quad \mathcal{N}^\nu_{\varrho_N}(A_2) \left(\bar{y}, \eta; \tau', \frac{1-\tau}{1-\tau'} r \right) d\tau' \left| \frac{dr}{r} d\tau \right|^{\frac{1}{2}} \\ &= \left(\mathcal{N}^\nu_{\varrho_N}(A_1) \mathcal{N}^\nu_{\varrho_N}(A_2) \right)(\bar{y}, \eta; \tau, r), \end{aligned}$$

where for the last equality we have used the fact that with respect to the coordinates $(\tau, r) \in [-1, 1] \times \overline{\mathbb{R}}_+$ on $M^2_{b,c}$ the lifted kernel $\kappa = \widehat{\kappa}(\tau, r) \left| \frac{dr}{r} d\tau \right|^{\frac{1}{2}}$ of the composition of two b-c-operators a_1 and a_2 is given by

$$\widehat{\kappa}(\tau, r) = \sqrt{2} \int_{-1}^{1} \frac{1}{1 - \tau\tau'} \widehat{\kappa}_{a_1} \left(\frac{\tau - \tau'}{1 - \tau\tau'}, \frac{1 - \tau\tau'}{1 - \tau'} r \right) \widehat{\kappa}_{a_2} \left(\tau', \frac{1-\tau}{1-\tau'} r \right) d\tau'.$$

$\square$

PROPOSITION 4.5.2. *Let $A \in \Psi_0^m(X; {}^0\Omega^{1/2})$ be arbitrary. Then we have for all $\eta \in T^* \partial X \setminus \{0\}$*

(4.32) $$\mathcal{N}^\nu_{\varrho_N}(A^\sharp)(\eta) = \mathcal{N}^\nu_{\varrho_N}(A)(\eta)^\sharp \in \Psi_{b,c}^{m,m}(M; {}^{b,c}\Omega^{1/2}),$$

where $(\cdot)^\sharp$ stands for the formal adjoint of an operator.

Proof: A straight-forward computation shows that with respect to local coordinates (τ, U, r, y') the lifted Schwartz kernel $\kappa_{A^\sharp}$ of $A^\sharp$ satisfies

$$\kappa_{A^\sharp} = \overline{\widehat{\kappa}}_A(-\tau, -U, r, y' + rU) \left| d\tau \, dU \frac{dr}{r^n} dy' \right|^{\frac{1}{2}},$$

hence, by (4.4) and another straight-forward computation

$$\mathcal{N}^\nu_{\varrho_N}(A^\sharp)(y, \eta; \tau, r) = \overline{\mathcal{N}^\nu_{\varrho_N}(A)}(y, \eta; -\tau, r) = \mathcal{N}^\nu_{\varrho_N}(A)(y, \eta)^\sharp(\tau, r).$$

$\square$

For the next proposition, let $\varrho_0 : M \to \overline{\mathbb{R}}_+ : z \mapsto z$ be a defining function for the component $\{0\}$ of the boundary of M.

PROPOSITION 4.5.3. *Let $A \in \Psi_0^m(X; {}^0\Omega^{1/2})$ be arbitrary. Then we have for all $z \in \mathbb{C}$ and all $\eta \in T^*\partial X \setminus \{0\}$*

$$\mathcal{N}^\nu_{\varrho_N}(\varrho_N^{-z} A \varrho_N^z)(\eta) = \varrho_0^{-z} \mathcal{N}^\nu_{\varrho_N}(A)(\eta) \varrho_0^z \in \Psi_{b,c}^{m,m}(M; {}^{b,c}\Omega^{1/2}).$$

Proof: It suffices to note that with respect to local coordinates as in (1.8) the lifted Schwartz kernel κ_B of $B = \varrho_N^{-z} A \varrho_N^z$ is given by $\kappa_B = \left(\frac{1+\tau}{1-\tau} \right)^{-z} \kappa_A$ and then to apply this to (4.4) and (4.5). $\square$

Consequently, we obtain for $A \in \Psi_0^{m,k}(X; {}^0\Omega^{1/2})$

$$\mathcal{N}_{\varrho_N}^\nu(\varrho_N^k A)(\eta) = \varrho_0^k \mathcal{N}_{\varrho_N}^\nu(A\varrho_N^k)(\eta)\varrho_0^{-k}, \tag{4.33}$$

and we could define the reduced normal operator $\mathcal{N}^{(k)}(A)$ of $A \in \Psi_0^{m,k}(X; {}^0\Omega^{1/2})$ either by $\mathcal{N}_{\varrho_N}^\nu(\varrho_N^k A)$ or $\mathcal{N}_{\varrho_N}^\nu(A\varrho_N^k)$. Note that in particular the invertibility of $\mathcal{N}^{(k)}(A)$ does not depend on the choice of the first or the second variant. However, the multiplicativity (4.31) of the reduced normal operator is lost, and we do not want to enter into the easy but clumsy details.

Recall that $A \in \Psi_0^m(X; {}^0\Omega^{1/2})$ is called *elliptic* if ${}^0\sigma^{(m)}(A)(\zeta) \neq 0$ for all $\zeta \in {}^0T^*X \setminus \{0\}$. We are going to show that the reduced normal operator of an elliptic 0-operator is a smooth family of Fredholm operators between appropriate weighted Sobolev spaces on M provided we can find the weight $\mathfrak{a} \in \mathbb{R}$ uniformly in the parameter $q \in \partial X$.

PROPOSITION 4.5.4. *Let $A \in \Psi_0^m(X; {}^0\Omega^{1/2})$ be elliptic and assume that there exists $\mathfrak{a}_0 \in \mathbb{R}$ such that*

$$I_A(\xi - i(\mathfrak{a}_0 + \frac{n-1}{2}))(q) \neq 0 \text{ for all } \xi \in \mathbb{R} \text{ and all } q \in \partial X. \tag{4.34}$$

Then, for any $\mathfrak{a}_1 \in \mathbb{R}$, the reduced normal operator $\mathcal{N}_{\varrho_N}^\nu(A)$ is a smooth family

$$\mathcal{N}_{\varrho_N}^\nu(A) : T^*\partial X \setminus \{0\} \longrightarrow \mathcal{L}(\varrho_0^{\mathfrak{a}_0}\varrho_1^{\mathfrak{a}_1} H_{b,c}^s(M, {}^{b,c}\Omega^{1/2}), \varrho_0^{\mathfrak{a}_0}\varrho_1^{\mathfrak{a}_1-m} H_{b,c}^{s-m}(M, {}^{b,c}\Omega^{1/2}))$$

of Fredholm operators.

Proof: This is just a combination of Theorem 3.6.1 and Proposition 4.4.4. □

Note that for each fixed $q \in \partial X$, there always exists a discrete set $D(q) \subseteq \mathbb{C}$ such that $I_A(w) \neq 0$ for all $w \notin D(q)$. To find a weight-line $\Gamma_{\mathfrak{a}_0} := \{\mathrm{Im}(w) = -\mathfrak{a}_0\}$ with $\Gamma_{\mathfrak{a}_0} \cap \bigcup_{q \in \partial X} D(q) = \emptyset$ is a serious restriction; however, it is often satisfied in applications in geometry [54, 55, 58].

4.6. The case of 0-differential operators

Let $P \in \mathrm{Diff}_0^m(X, {}^0\Omega^{1/2})$ be a 0-differential operator of order m, i.e. with respect to local coordinates $(x, y) : X \supseteq V \to \overline{\mathbb{R}}_+ \times \mathbb{R}_y^{n-1}$ as in Lemma 1.1.2 we find $a_{j,\alpha} \in \mathcal{C}^\infty(\overline{\mathbb{R}}_+ \times \mathbb{R}_y^{n-1})$ such that we have for $f = \widehat{f}(x,y)\left|\frac{dx}{x^n}dy\right|^{\frac{1}{2}} \in \mathcal{C}^\infty(X, {}^0\Omega^{1/2})$

$$Pf(x,y) = \sum_{j+|\alpha| \leq m} a_{j,\alpha}(x,y)(x\partial_x)^j(x\partial_y)^\alpha \widehat{f}(x,y)\left|\frac{dx}{x^n}dy\right|^{\frac{1}{2}}.$$

Moreover, note that the choice of local coordinates (x, y) also yields trivializations ${}^0T^*X|_V \cong \overline{\mathbb{R}}_+ \times \mathbb{R}_y^{n-1} \times \mathbb{R}_\xi \times \mathbb{R}_\eta^{n-1}$ as well as $T^*\partial X|_{\partial X \cap V} \cong \mathbb{R}_y^{n-1} \times \mathbb{R}_\eta^{n-1}$. We summarize the expressions for the various symbols of the 0-differential operator P with respect to these local coordinates in the next Proposition.

PROPOSITION 4.6.1.
$$^0\sigma^{(m)}(P)(x,y,\xi,\eta) = i^m \sum_{j+|\alpha|=m} a_{j,\alpha}(x,y)\xi^j \eta^\alpha,$$

$$(4.35) \quad \left[\mathcal{N}^\nu_{\varrho_N}(P)(y,\eta)\right] f(x) = \sum_{j+|\alpha|\leq m} a_{j,\alpha}(0,y) i^{|\alpha|} x^{|\alpha|} \eta^\alpha (x\partial_x)^j \widehat{f}(x) \left|\frac{dx}{x}\right|^{\frac{1}{2}},$$

$$I_P(w)(y) = \sum_{j=0}^m a_{j,0}(0,y) i^j w^j = I_b\left(\mathcal{N}^\nu_{\varrho_N}(P)(y,\eta)\right)(w),$$

$$^{b,c}\sigma^{(m,m)}(\mathcal{N}^\nu_{\varrho_N}(P)(y,\eta))(z,\xi) = i^m a_{m,0}(0,y)\xi^m, \text{ and}$$

$$I_c^{(m)}(\mathcal{N}^\nu_{\varrho_N}(P)(y,\eta))(\xi) = i^m \sum_{j+|\alpha|=m} a_{j,\alpha}(0,y)\xi^j \eta^\alpha.$$

Note that in (4.35) we used the identification of $\overline{\mathbb{R}}_+$ with $[0,1) \subseteq M$ given by the radial compactification (3.1). Thus, for 0-differential operators the definition of the reduced normal operator coincides with the one given in [**54, 55, 58**] after using the scale invariance.

4.7. General bundles

The definition of the reduced normal operator naturally extends to the case of 0-operators acting between sections of finite dimensional vector bundles $E, F \to X$. Indeed, for $A \in \Psi_0^{m,k,\gamma}(X; E, F)$, the reduced normal operator is a smooth family

$$\mathcal{N}^\nu_{\varrho_N}(A) : T^*\partial X \setminus \{0\} \ni \eta \longmapsto \Psi_{b,c}^{m,m}(M; E_{\pi(\eta)}, F_{\pi(\eta)})$$

of b-c-operators on M acting on sections of the trivial bundles $E_{\pi(\eta)}$ and $F_{\pi(\eta)}$. The results of this section remain true with the obvious changes.

CHAPTER 5

Weighted 0-Sobolev spaces

Let $L^2(X, {}^0\Omega^{1/2})$ be the Hilbert space completion of $\dot{\mathcal{C}}^\infty(X, {}^0\Omega^{1/2})$ with respect to the scalar product
$$<f, g>_{L^2(X, {}^0\Omega^{1/2})} := \int_X f\bar{g}.$$

5.1. Boundedness of 0-operators of order 0 on L^2-spaces

The basic fact that, for appropriate weights γ, 0-pseudodifferential operators in $\Psi_0^{0,0,\gamma}(X; {}^0\Omega^{1/2})$ act as bounded operators on $L^2(X, {}^0\Omega^{1/2})$ can be reduced to the case $\Psi_0^{-\infty,0,\gamma}(X; {}^0\Omega^{1/2})$ by Proposition 2.3.8. So let us start with that case.

PROPOSITION 5.1.1. *Let $\gamma = (\gamma_T, \gamma_B, \gamma_F)$ be a weight system for X with $\gamma_T > -1/2$, $\gamma_B > -1/2$, and $\gamma_F > -1/2$. Then each $A \in \Psi_0^{-\infty,0,\gamma}(X; {}^0\Omega^{1/2})$ extends from $A : \dot{\mathcal{C}}^\infty(X, {}^0\Omega^{1/2}) \to \mathcal{C}^{-\infty}(X, {}^0\Omega^{1/2})$ to a bounded linear operator $A : L^2(X, {}^0\Omega^{1/2}) \to L^2(X, {}^0\Omega^{1/2})$.*

Proof: If $\operatorname{supp}\kappa_A \cap \operatorname{ff}^{0,e} = \emptyset$, then κ_A is the lift of a kernel in the Hilbert space $L^2(X^2, {}^0\Omega^{1/2} \boxtimes {}^0\Omega^{1/2})$, thus, A is a Hilbert-Schmidt operator on $L^2(X, {}^0\Omega^{1/2})$, hence, it is bounded. Therefore, we can assume that κ_A is supported close to the front face $\operatorname{ff}^{0,e}$, and we can use polar coordinates as introduced in (1.7) corresponding to local coordinates $(x, y) : X \supseteq V \to \overline{\mathbb{R}}_+ \times \mathbb{R}_y^{n-1}$ near ∂X. Let $\varepsilon > 0$ be with $\gamma_O > -1/2 + \varepsilon$, $O = B, F, T$, and let $f_j = \hat{f}_j(x, y) \left|\frac{dx}{x^n} dy\right|^{\frac{1}{2}} \in \dot{\mathcal{C}}^\infty(X, {}^0\Omega^{1/2})$ be arbitrary. If
$$\kappa_A = \hat{\kappa}_A(\tau, \vartheta, \varrho, y') \left| d\tau \, d\vartheta \frac{d\varrho}{\varrho^n} dy' \right|^{\frac{1}{2}} \in \mathcal{K}_0^{-\infty,0,\gamma}(X_{0,e}^2, KD_{0,e}^{1/2})$$

is the lifted Schwartz kernel of A, we have in particular,

(5.1) $\qquad |\hat{\kappa}_A(\tau, \vartheta, \varrho, y')| \leq \operatorname{const}(1+\tau)^{\gamma_T}(1-\tau)^{\gamma_B}\vartheta_0^{\gamma_F},$

and we can assume that $\hat{\kappa}_A$ vanishes for large ϱ and $|y'|$. Since we have up to a positive constant
$$(\beta_{0,e}^2)^* \left| \frac{dx}{x^n} dy \frac{dx'}{x'^n} dy' \right|^{\frac{1}{2}} = \left| \frac{d\tau}{(1-\tau^2)^n} \frac{d\vartheta}{\vartheta_0^{2n-1}} \frac{d\varrho}{\varrho^n} dy' \right|^{\frac{1}{2}}$$

we obtain for some large $R > 0$

$$<Af_1, f_2> = \int_{|y'|\leq R} \int_0^R \int_{S_+^{n-1}} \int_{-1}^1 \hat{\kappa}_A(\tau, \vartheta, \varrho, y') \vartheta_0^{(2n-1)/2}(1-\tau^2)^{n/2}$$
$$\hat{f}_1(\varrho\vartheta_0(1-\tau)/2, y') \cdot \overline{\hat{f}_2(\varrho\vartheta_0(1+\tau)/2, y' + \varrho\vartheta')} \frac{d\tau}{(1-\tau^2)^n} \frac{d\vartheta}{\vartheta_0^{2n-1}} \frac{d\varrho}{\varrho^n} dy'.$$

Therefore, (5.1) together with Cauchy-Schwartz inequality yields

$$|<Af_1, f_2>|^2$$
$$\leq \text{const} \left(\int_{S_+^{n-1}} \int_{-1}^{1} \int_{|y'|\leq R} \int_0^R (1+\tau)^{2\varepsilon-1}(1-\tau)^{2\gamma_B-n+1-2\varepsilon} \vartheta_0^{2\varepsilon-n} \right.$$
$$\left. \left|\widehat{f_1}(\varrho\vartheta_0(1-\tau)/2, y')\right|^2 \frac{d\varrho}{\varrho^n} dy' d\tau d\vartheta \right) \times$$
$$\times \left(\int_{S_+^{n-1}} \int_{-1}^{1} \int_{|y'|\leq R} \int_0^R (1+\tau)^{2\gamma_T-n+1-2\varepsilon}(1-\tau)^{2\varepsilon-1} \vartheta_0^{2\gamma_F-2\varepsilon-n+1} \right.$$
$$\left. \left|\widehat{f_2}(\varrho\vartheta_0(1+\tau)/2, y' + \varrho\vartheta')\right|^2 \frac{d\varrho}{\varrho^n} dy' d\tau d\vartheta \right)$$
$$\leq \text{const} \left(\int_{S_+^{n-1}} \int_{-1}^{1} (1+\tau)^{2\varepsilon-1}(1-\tau)^{2(\gamma_B-\varepsilon)} \vartheta_0^{2\varepsilon-1} \right.$$
$$\left(\int_{\mathbb{R}_{y'}^{n-1}} \int_0^\infty \left|\widehat{f_1}(x',y')\right|^2 \frac{dx'}{x'^n} dy' \right) d\tau d\vartheta \right) \times$$
$$\times \left(\int_{S_+^{n-1}} \int_{-1}^{1} (1+\tau)^{2(\gamma_T-\varepsilon)}(1-\tau)^{2\varepsilon-1} \vartheta_0^{2(\gamma_F-\varepsilon)} \right.$$
$$\left(\int_{\mathbb{R}_y^{n-1}} \int_0^\infty \left|\widehat{f_2}(x,y)\right|^2 \frac{dx}{x^n} dy \right) d\tau d\vartheta \right)$$
$$\leq \text{const} \|f_1\|^2_{L^2(X, {}^0\Omega^{1/2})} \|f_2\|^2_{L^2(X, {}^0\Omega^{1/2})}.$$

Since $\dot{\mathcal{C}}^\infty(X, {}^0\Omega^{1/2})$ is dense in $L^2(X, {}^0\Omega^{1/2})$, this implies

$$\|Af\|_{L^2(X, {}^0\Omega^{1/2})} \leq \text{const} \|f\|_{L^2(X, {}^0\Omega^{1/2})}$$

with a constant independent of f. $\square$

THEOREM 5.1.2. *Let $\gamma = (\gamma_T, \gamma_B, \gamma_F)$ be an arbitrary weight system for X with $\gamma_T > -1/2$, $\gamma_B > -1/2$, and $\gamma_F > -1/2$. Then $A \in \Psi_0^{0,0,\gamma}(X; {}^0\Omega^{1/2})$ extends from $A : \dot{\mathcal{C}}^\infty(X, {}^0\Omega^{1/2}) \to \mathcal{C}^{-\infty}(X, {}^0\Omega^{1/2})$ to a bounded linear operator $A : L^2(X, {}^0\Omega^{1/2}) \to L^2(X, {}^0\Omega^{1/2})$.*

Proof: By Proposition 5.1.1 we can assume $A \in \Psi_0^0(X; {}^0\Omega^{1/2})$, thus, there exists $M > 0$ with ${}^0\sigma^{(0)}(M\text{id} - A^\sharp A) > 0$, hence Proposition 2.3.8 gives $B \in \Psi_0^0(X; {}^0\Omega^{1/2})$ and $R \in \Psi_0^{-\infty}(X; {}^0\Omega^{1/2})$ with $M\text{id} - A^\sharp A = B^\sharp B + R$. Therefore, we obtain for every $f \in \dot{\mathcal{C}}^\infty(X, {}^0\Omega^{1/2})$

$$\begin{aligned}
\|Af\|^2_{L^2(X, {}^0\Omega^{1/2})} &= <A^\sharp A f, f>_{L^2(X, {}^0\Omega^{1/2})} \\
&= M\|f\|^2_{L^2(X, {}^0\Omega^{1/2})} - \|Bf\|^2_{L^2(X, {}^0\Omega^{1/2})} - <Rf, f>_{L^2(X, {}^0\Omega^{1/2})} \\
&\leq \left(M + \|R\|_{\mathcal{L}(L^2(X, {}^0\Omega^{1/2}))}\right) \|f\|^2_{L^2(X, {}^0\Omega^{1/2})}
\end{aligned}$$

by Proposition 5.1.1 which completes the proof. $\square$

5.2. WEIGHTED 0-SOBOLEV SPACES

Later on we will also need a weighted version of $L^2(X, {}^0\Omega^{1/2})$. Let $\mathfrak{a} \in \mathbb{R}$ be arbitrary, and define

(5.2) $\quad \varrho_N^{\mathfrak{a}} L^2(X, {}^0\Omega^{1/2}) := \left\{ f \in \mathcal{C}^{-\infty}(X, {}^0\Omega^{1/2}) : \varrho_N^{-\mathfrak{a}} f \in L^2(X, {}^0\Omega^{1/2}) \right\}.$

The space $\varrho_N^{\mathfrak{a}} L^2(X, {}^0\Omega^{1/2})$ clearly becomes a Hilbert space with respect to the scalar product

(5.3)
$$<f, g>_{\varrho_N^{\mathfrak{a}} L^2(X, {}^0\Omega^{1/2})} := <\varrho_N^{-\mathfrak{a}} f, \varrho_N^{-\mathfrak{a}} g>_{L^2(X, {}^0\Omega^{1/2})} \text{ for } f, g \in \varrho_N^{\mathfrak{a}} L^2(X, {}^0\Omega^{1/2})$$

depending on the choice of the boundary defining function ϱ_N. Moreover, the space $\dot{\mathcal{C}}^{\infty}(X, {}^0\Omega^{1/2})$ is dense in $\varrho_N^{\mathfrak{a}} L^2(X, {}^0\Omega^{1/2})$.

THEOREM 5.1.3. *Let $\mathfrak{a} \in \mathbb{R}$ be arbitrary, and $\gamma = (\gamma_T, \gamma_B, \gamma_F)$ be a weight system with $\gamma_T > \mathfrak{a} - 1/2$, $\gamma_B > -\mathfrak{a} - 1/2$, and $\gamma_F > -1/2$. Then each 0-operator $A \in \Psi_0^{0,0,\gamma}(X; {}^0\Omega^{1/2})$ extends from $\dot{\mathcal{C}}^{\infty}(X, {}^0\Omega^{1/2})$ to a bounded operator $A : \varrho_N^{\mathfrak{a}} L^2(X, {}^0\Omega^{1/2}) \to \varrho_N^{\mathfrak{a}} L^2(X, {}^0\Omega^{1/2})$.*

Proof: We have $\varrho_N^{-\mathfrak{a}} A \varrho_N^{\mathfrak{a}} \in \Psi_0^{0,0,\gamma-\mathfrak{a}}(X; {}^0\Omega^{1/2})$ with $(\gamma_{-\mathfrak{a}})_T = \gamma_T - \mathfrak{a} > -1/2$, $(\gamma_{-\mathfrak{a}})_B = \gamma_B + \mathfrak{a} > -1/2$, and $(\gamma_{-\mathfrak{a}})_F = \gamma_F > -1/2$ by Proposition 2.5.1; thus, $\varrho_N^{-\mathfrak{a}} A \varrho_N^{\mathfrak{a}} : L^2(X, {}^0\Omega^{1/2}) \to L^2(X, {}^0\Omega^{1/2})$ is bounded which implies the boundedness of $A = \varrho_N^{\mathfrak{a}}(\varrho_N^{-\mathfrak{a}} A \varrho_N^{\mathfrak{a}})\varrho_N^{-\mathfrak{a}} : \varrho_N^{\mathfrak{a}} L^2(X, {}^0\Omega^{1/2}) \to \varrho_N^{\mathfrak{a}} L^2(X, {}^0\Omega^{1/2})$, and completes the proof. $\square$

Since $\dot{\mathcal{C}}^{\infty}(X, {}^0\Omega^{1/2})$ is dense in the Hilbert space $\varrho_N^{\mathfrak{a}} L^2(X, {}^0\Omega^{1/2})$ for any $\mathfrak{a} \in \mathbb{R}$, the natural map $\Psi_0^0(X; {}^0\Omega^{1/2}) \to \mathcal{L}(\varrho_N^{\mathfrak{a}} L^2(X, {}^0\Omega^{1/2}))$ is one-to-one, and we can think of $\Psi_0^0(X; {}^0\Omega^{1/2})$ as a subalgebra of the C^*-algebra $\mathcal{L}(\varrho_N^{\mathfrak{a}} L^2(X, {}^0\Omega^{1/2}))$. Note that besides the case $\mathfrak{a} = 0$, the C^*-structure on $\mathcal{L}(\varrho_N^{\mathfrak{a}} L^2(X, {}^0\Omega^{1/2}))$ depends on the choice of the boundary defining function $\varrho_N : X \to \overline{\mathbb{R}}_+$.

LEMMA 5.1.4. *For any weight $\mathfrak{a} \in \mathbb{R}$, the algebra $\Psi_0^0(X; {}^0\Omega^{1/2})$ is a symmetric subalgebra of the C^*-algebra $\mathcal{L}(\varrho_N^{\mathfrak{a}} L^2(X, {}^0\Omega^{1/2}))$.*

Proof: By Lemma 2.3.3, the Hilbert space adjoint of $A \in \Psi_0^0(X; {}^0\Omega^{1/2})$ is given by $A^{*\mathfrak{a}} = \varrho_N^{\mathfrak{a}}(\varrho_N^{-\mathfrak{a}} A \varrho_N^{\mathfrak{a}})^{\sharp} \varrho_N^{-\mathfrak{a}} \in \Psi_0^0(X; {}^0\Omega^{1/2})$. $\square$

5.2. Weighted 0-Sobolev spaces

The construction of a scale of weighted 0-Sobolev spaces on X mainly uses the symbolic properties of the 0-calculus. A similar construction for double-edge Sobolev spaces has been carried out in [**44**] in detail, whereas a thorough treatment of weighted b-Sobolev spaces on manifolds with corners can be found in [**51**]. Thus, we sketch the essential steps and list the relevant properties only, referring to [**44**, Section 4] and [**51**] for complete proofs.

DEFINITION 5.2.1. For arbitrary $\mathfrak{a} \in \mathbb{R}$ and $s \in \mathbb{R}$, the space
(5.4)
$$\varrho_N^{\mathfrak{a}} H_0^s(X, {}^0\Omega^{1/2}) := \left\{ f \in \mathcal{C}^{-\infty}(X, {}^0\Omega^{1/2}) : \Psi_0^{s,\mathfrak{a}}(X; {}^0\Omega^{1/2}) f \subseteq L^2(X, {}^0\Omega^{1/2}) \right\}$$

is called the *0-Sobolev space of order s and weight $\mathfrak{a}$*.

As an immediate consequence of this definition we obtain

(5.5) $\quad \varrho_N^{\mathfrak{a}} H_0^s(X, {}^0\Omega^{1/2}) \subseteq \varrho^{\mathfrak{a}_1} H_0^{s_1}(X, {}^0\Omega^{1/2})$ provided $s - s_1 \in \mathbb{N}_0$ and $\mathfrak{a} - \mathfrak{a}_1 \in \mathbb{N}_0$, and $A\left(\varrho_N^{\mathfrak{a}} H_0^s(X, {}^0\Omega^{1/2})\right) \subseteq \varrho_N^{\mathfrak{a}-k} H_0^{s-m}(X, {}^0\Omega^{1/2})$ for all $A \in \Psi_0^{m,k}(X; {}^0\Omega^{1/2})$ by (2.11). With a little more effort one can actually prove a more precise characterization, namely

(5.6) $\qquad \varrho_N^{\mathfrak{a}} H_0^s(X, {}^0\Omega^{1/2}) \subseteq \varrho_N^{\mathfrak{a}_1} H_0^{s_1}(X, {}^0\Omega^{1/2}) \iff s \geq s_1$ and $\mathfrak{a} \geq \mathfrak{a}_1$.

Let us summarize the main properties of the spaces $\varrho_N^{\mathfrak{a}} H_0^s(X, {}^0\Omega^{1/2})$ in the next theorem.

THEOREM 5.2.2. (a) *The spaces $\varrho_N^{\mathfrak{a}} H_0^s(X, {}^0\Omega^{1/2})$ are Hilbertable topological vector spaces.*
(b) *For all $\mathfrak{a}, s \in \mathbb{R}$, the space $\dot{\mathcal{C}}^\infty(X, {}^0\Omega^{1/2})$ is dense in $\varrho_N^{\mathfrak{a}} H_0^s(X, {}^0\Omega^{1/2})$.*
(c) *For $s \geq s_1$ and $\mathfrak{a} \geq \mathfrak{a}_1$, the canonical embedding*

$$\varrho_N^{\mathfrak{a}} H_0^s(X, {}^0\Omega^{1/2}) \hookrightarrow \varrho_N^{\mathfrak{a}_1} H_0^{s_1}(X, {}^0\Omega^{1/2})$$

is continuous; it is compact provided $s > s_1$ and $\mathfrak{a} > \mathfrak{a}_1$.
(d) *Let $\mathfrak{a}, \mathfrak{a}', k, m, s, s' \in \mathbb{R}$ and γ be a weight system for X with $m \leq s - s'$, $k \leq \mathfrak{a} - \mathfrak{a}'$, $\gamma_T > \mathfrak{a}' - 1/2$, $\gamma_B > -\mathfrak{a} - 1/2$, and $\gamma_F > -1/2$. Then any $A \in \Psi_0^{m,k,\gamma}(X; {}^0\Omega^{1/2})$ induces a bounded operator*

$$A : \varrho_N^{\mathfrak{a}} H_0^s(X, {}^0\Omega^{1/2}) \longrightarrow \varrho_N^{\mathfrak{a}'} H_0^{s'}(X, {}^0\Omega^{1/2}).$$

This operator is compact provided we have additionally $m < s - s'$ and $k < \mathfrak{a} - \mathfrak{a}'$.

5.3. General bundles

The results of this section extend directly to sections with values in general, finite-dimensional complex vector bundles over X. However, for the definition of the scalar product in the space $H_0^0(X, E) = L^2(X, E)$ we have to fix a hermitian metric on E and a 0-density on X. Instead, we could consider the bundle $E \otimes {}^0\Omega^{1/2}$.

CHAPTER 6

Fredholm theory for 0-pseudodifferential operators

6.1. Symbol reproducing families

As in the case of classical pseudodifferential operators on closed manifolds, the homogeneous principal symbol of a 0-pseudodifferential operators $A \in \Psi_0^0(X; {}^0\Omega^{1/2})$ can be recovered from the action of A on a certain system of functions. The construction of these *symbol reproducing families* goes back at least as far as [20]. For the sake of completeness, we are going to sketch the construction of this *oscillatory testing*. Let $\pi : {}^0S^*X \to X$ be the canonical projection.

PROPOSITION 6.1.1. *Let $\omega_\sigma \in {}^0S^*X$ be arbitrary. Then for each $t > 0$, there exists $g_t^{\omega_\sigma} \in \dot{\mathcal{C}}^\infty(X, {}^0\Omega^{1/2})$ with $\|g_t^{\omega_\sigma}\|_{L^2(X, {}^0\Omega^{1/2})} = 1$, $g_t^{\omega_\sigma} \xrightarrow{t \to \infty} 0$ weakly in $L^2(X, {}^0\Omega^{1/2})$, and $\lim_{t \to \infty} \|Ag_t^{\omega_\sigma}\| = |{}^0\sigma^{(0)}(A)(\omega_\sigma)|$ for all $A \in \Psi_0^0(X; {}^0\Omega^{1/2})$.*

Proof: For $\pi(\omega_\sigma) \notin \partial X$ we use the standard construction for closed manifolds; so assume $\pi(\omega_\sigma) \in \partial X$. Let $(x, y) : X \supseteq U \to \overline{\mathbb{R}}_+ \times \mathbb{R}_y^{n-1}$ be local coordinates near $\pi(\omega_\sigma)$, and let

$$(x, y, \xi, \eta) : {}^0T^*X \supseteq {}^0T^*X|_U \longrightarrow \overline{\mathbb{R}}_+ \times \mathbb{R}_y^{n-1} \times \mathbb{R}_\xi \times \mathbb{R}_\eta^{n-1}$$

be the corresponding trivialization of ${}^0T^*X$. Let $(0, 0, \xi^{(0)}, \eta^{(0)}) \in {}^0T^*X \setminus \{0\}$ be a representative of ω_σ with respect to these coordinates. Choose $v_0 \in \mathcal{C}_c^\infty(\mathbb{R}_+)$ with $\int_0^\infty |v_0(x)|^2 \frac{dx}{x} = 1$, $v_1 \in \mathcal{C}_c^\infty(\mathbb{R}_y^{n-1})$ with $\int_{\mathbb{R}_y^{n-1}} |v_1(y)|^2 dy = 1$, and define for $t > 0$ the section $g_t^{\omega_\sigma} \in \dot{\mathcal{C}}^\infty(X, {}^0\Omega^{1/2})$ in local coordinates by

$$g_t^{\omega_\sigma} : (x, y) \longmapsto \sqrt{2} t^{\frac{n}{2}} x^{-\frac{n-1}{2}} x^{2it^2 \xi^{(0)}} e^{it^3 y \eta^{(0)}} v_0\left((tx)^t\right) v_1(ty) \left|\frac{dx}{x^n} dy\right|^{\frac{1}{2}}.$$

Using Lebesgue's theorem twice, we see that $g_t^{\omega_\sigma}$ has the desired properties. $\square$

We are now going to consider the reduced normal operator.

PROPOSITION 6.1.2. *Let $\eta \in T^*\partial X \setminus \{0\}$ and $f \in \dot{\mathcal{C}}^\infty(M, {}^{b,c}\Omega^{1/2})$ be arbitrary with $\|f\|_{L^2(M, {}^{b,c}\Omega^{1/2})} = 1$ and $f(z) = 0$ for $1 - \delta \leq z \leq 1$ and some $\delta > 0$. Then, for each $t > 0$, there exists $u_t(f) \in \dot{\mathcal{C}}^\infty(X, {}^0\Omega^{1/2})$ with $\|u_t(f)\|_{L^2(X, {}^0\Omega^{1/2})} = 1$, $u_t(f) \xrightarrow{t \to \infty} 0$ weakly in $L^2(X, {}^0\Omega^{1/2})$, and, for all $A \in \Psi_0^0(X; {}^0\Omega^{1/2})$,*

$$\lim_{t \to \infty} \|Au_t(f)\|_{L^2(X, {}^0\Omega^{1/2})} = \|\mathcal{N}_{\varrho N}^\nu(A)(\eta)f\|_{L^2(M, {}^{b,c}\Omega^{1/2})}.$$

Proof: Identifying $M \setminus \{1\}$ with $\overline{\mathbb{R}}_+$, we can assume that we have $f = \widehat{f}(x) \left|\frac{dx}{x}\right|^{\frac{1}{2}}$ with some $\widehat{f} \in \dot{\mathcal{C}}_c^\infty(\overline{\mathbb{R}}_+)$ and $\int_0^\infty |\widehat{f}(x)|^2 \frac{dx}{x} = 1$. Choose now $\widehat{g} \in \mathcal{C}_c^\infty(\mathbb{R}_y^{n-1})$ with $\int_{\mathbb{R}_y^{n-1}} |\widehat{g}(y)|^2 dy = 1$, and define up to the choice of local coordinates as in the proof of Proposition 6.1.1

$$u_t(f) : (x,y) \longmapsto t^{\frac{n-1}{2}} x^{\frac{n-1}{2}} e^{it^2 y\eta} \widehat{f}(t^2 x) \widehat{g}(ty) \left| \frac{dx}{x^n} dy \right|^{\frac{1}{2}}.$$

As in the proof of Proposition 6.1.1, an application of Lebesgue's theorem shows that $u_t(f)$ has the desired properties. □

COROLLARY 6.1.3. *Let $A \in \Psi_0^0(X; {}^0\Omega^{1/2})$ be arbitrary. If A induces a Fredholm operator $A : L^2(X, {}^0\Omega^{1/2}) \to L^2(X, {}^0\Omega^{1/2})$, then the reduced normal operator $\mathcal{N}_{\varrho_N}^\nu(A)(\eta) : L^2(M, {}^{b,c}\Omega^{1/2}) \to L^2(M, {}^{b,c}\Omega^{1/2})$ is one-to-one for all $\eta \in T^*\partial X \setminus \{0\}$.*

Proof: Since A is Fredholm, there exists $B \in \mathcal{L}(L^2(X, {}^0\Omega^{1/2}))$ such that the operator $K := \text{id} - BA \in \mathcal{K}(L^2(X, {}^0\Omega^{1/2}))$ is compact. If $0 \neq f \in L^2(M, {}^{b,c}\Omega^{1/2})$ satisfies $\mathcal{N}_{\varrho_N}^\nu(A)(\eta)f = 0$, then there exists a sequence $f_j \in \dot{\mathcal{C}}^\infty(M, {}^{b,c}\Omega^{1/2})$ with $f_j(z) = 0$ for $1 - \delta_j \leq z \leq 1$ and $f_j \to f$ in $L^2(M, {}^{b,c}\Omega^{1/2})$. By Proposition 6.1.2 we obtain a sequence $h_k \in \dot{\mathcal{C}}^\infty(X, {}^0\Omega^{1/2})$ with $\|h_k\|_{L^2(X, {}^0\Omega^{1/2})} = 1$, $\|Ah_k\|_{L^2(X, {}^0\Omega^{1/2})} \to \|\mathcal{N}_{\varrho_N}^\nu(A)(\eta)f\|_{L^2(M, {}^{b,c}\Omega^{1/2})} = 0$, and $\|Kh_k\|_{L^2(X, {}^0\Omega^{1/2})} \to 0$ as $k \to \infty$. Thus, we get a contradiction because of

$$1 = \|h_k\|_{L^2(X, {}^0\Omega^{1/2})} \leq \underbrace{\|B\|_{\mathcal{L}(L^2(X, {}^0\Omega^{1/2}))} \|Ah_k\|_{L^2(X, {}^0\Omega^{1/2})} + \|Kh_k\|_{L^2(X, {}^0\Omega^{1/2})}}_{\xrightarrow{k \to \infty} 0}.$$

□

Finally, we need a similar result for the indicial function of a 0-pseudodifferential operator.

PROPOSITION 6.1.4. *Let $q \in \partial X$ and $\xi \in \mathbb{R}$ be arbitrary. Then for $t > 0$, there exists $h_t \in \dot{\mathcal{C}}^\infty(X, {}^0\Omega^{1/2})$ with $\|h_t\|_{L^2(X, {}^0\Omega^{1/2})} = 1$, $h_t \xrightarrow{t \to \infty} 0$ weakly in $L^2(X, {}^0\Omega^{1/2})$, and, for all $A \in \Psi_0^0(X; {}^0\Omega^{1/2})$ and all $\eta \in T^*\partial X \setminus \{0\}$ with $\pi(\eta) = q$*

$$\lim_{t \to \infty} \|Ah_t\|_{L^2(X, {}^0\Omega^{1/2})} = \left| I_A(\xi - i\frac{n-1}{2})(q) \right| = \left| I_b \left(\mathcal{N}_{\varrho_N}^\nu(A)(\eta) \right)(\xi) \right|.$$

Proof: Let $(x,y) : X \supseteq U \to \overline{\mathbb{R}}_+ \times \mathbb{R}_y^{n-1}$ be local coordinates near $q \in \partial X$ with $(x,y)(q) = (0,0)$, and choose $v_0 \in \mathcal{C}_c^\infty((0,1))$ with $\int_0^\infty |v_0(x)|^2 \frac{dx}{x} = 1$ as well as $v_1 \in \mathcal{C}_c^\infty(\mathbb{R}_y^{n-1})$ with $\int_{\mathbb{R}_y^{n-1}} |v_1(y)|^2 dy = 1$. Then the smooth section

$$h_t : (x,y) \longmapsto t^{\frac{n-2}{2}} x^{\frac{n-1}{2}} x^{i\xi} v_0(x^{1/t}) v_1(ty) \left| \frac{dx}{x^n} dy \right|^{\frac{1}{2}}$$

has the desired properties. □

6.2. Characterization of Fredholm operators in $\Psi_0^0(X; {}^0\Omega^{1/2})$

Recall that each 0-operator $A \in \Psi_0^0(X; {}^0\Omega^{1/2})$ induces a bounded operator $A : L^2(X, {}^0\Omega^{1/2}) \to L^2(X, {}^0\Omega^{1/2})$ by Proposition 5.1.1.

THEOREM 6.2.1. *Let $A \in \Psi_0^0(X; {}^0\Omega^{1/2})$ be arbitrary. Then the bounded operator $A : L^2(X, {}^0\Omega^{1/2}) \to L^2(X, {}^0\Omega^{1/2})$ is Fredholm if and only if ${}^0\sigma^{(0)}(A)(\zeta) \neq 0$ for all $\zeta \in {}^0S^*X$, and $\mathcal{N}_{\varrho_N}^\nu(A)(\eta) : L^2(M, {}^{b,c}\Omega^{1/2}) \to L^2(M, {}^{b,c}\Omega^{1/2})$ is invertible for all $\eta \in T^*\partial X \setminus \{0\}$.*

6.2. CHARACTERIZATION OF FREDHOLM OPERATORS IN $\Psi_0^0(X; {}^0\Omega^{1/2})$

By the scale invariance of the reduced normal operator (Corollary 4.3.1), it is of course sufficient to check the invertibility of $\mathcal{N}_{\varrho_N}^\nu(A)(\eta)$ only for one representative $\eta \in T^*\partial X \setminus \{0\}$ for each class $[\eta] \in S^*\partial X$.

Proof: Let us start with the necessity of the conditions. Let $B \in \mathcal{L}(L^2(X, {}^0\Omega^{1/2}))$ be an operator such that $K := \mathrm{id} - BA : L^2(X, {}^0\Omega^{1/2}) \to L^2(X, {}^0\Omega^{1/2})$ is compact. For $\omega_\sigma \in {}^0S^*X$, let $g_t^{\omega_\sigma} \in \dot{\mathcal{C}}^\infty(X, {}^0\Omega^{1/2})$ be as in Proposition 6.1.1. Then we get

$$1 = \|g_t^{\omega_\sigma}\|_{L^2(X,{}^0\Omega^{1/2})} \leq \underbrace{\|B\|\, \|Ag_t^{\omega_\sigma}\|_{L^2(X,{}^0\Omega^{1/2})} + \|Kg_t^{\omega_\sigma}\|_{L^2(X,{}^0\Omega^{1/2})}}_{\xrightarrow{t \to \infty} \|B\| |{}^0\sigma^{(0)}(A)(\omega_\sigma)|},$$

hence, we have ${}^0\sigma^{(0)}(A)(\omega_\sigma) \neq 0$. By Proposition 4.5.2 and Corollary 6.1.3, we know already that $\mathcal{N}_{\varrho_N}^\nu(A)(\eta)$ and its Hilbert space adjoint $\mathcal{N}_{\varrho_N}^\nu(A)(\eta)^*$ are one-to-one; it remains to check that the range of $\mathcal{N}_{\varrho_N}^\nu(A)(\eta)$ is closed. Indeed, by Proposition 4.4.4, we have ${}^{b,c}\sigma^{(0,0)}(\mathcal{N}_{\varrho_N}^\nu(A)(\eta))(z,\xi) = {}^0\sigma^{(0)}(A)\left(j_{\varrho_N}(0,(\pi(\eta),\xi))\right) \neq 0$ for all $(z,\xi) \in {}^{b,c}T^*M \setminus \{0\}$ and $I_c^{(0)}(\mathcal{N}_{\varrho_N}^\nu(A)(\eta))(\xi) = {}^0\sigma^{(0)}(A)\left(j_{\varrho_N}(\eta,(\pi(\eta),\xi))\right) \neq 0$ for all $\xi \in \mathbb{R}$. Moreover, for $q \in \partial X$ and $\xi \in \mathbb{R}$, let h_t be as in Proposition 6.1.4; then we get

$$1 = \|h_t\|_{L^2(X,{}^0\Omega^{1/2})} \leq \underbrace{\|B\|\,\|Ah_t\|_{L^2(X,{}^0\Omega^{1/2})} + \|Kh_t\|_{L^2(X,{}^0\Omega^{1/2})}}_{\xrightarrow{t \to \infty} \|B\| |I_b(\mathcal{N}_{\varrho_N}^\nu(A)(\eta))(\xi)|},$$

thus, the reduced normal operator $\mathcal{N}_{\varrho_N}^\nu(A)(\eta) : L^2(M, {}^{b,c}\Omega^{1/2}) \to L^2(M, {}^{b,c}\Omega^{1/2})$ is Fredholm by Theorem 3.6.1, and its range is closed.

As for the sufficiency of the condition, first note that by Theorem 3.6.1 there exists $\gamma_{\mathrm{lb}} > -1/2$ and $\gamma_{\mathrm{rb}} > -1/2$ with $\mathcal{N}_{\varrho_N}^\nu(A)(\eta)^{-1} \in \Psi_{b,c}^{0,0,(\gamma_{\mathrm{lb}},\gamma_{\mathrm{rb}})}(M; {}^{b,c}\Omega^{1/2})$ for all $\eta \in T^*\partial X \setminus \{0\}$,

$$\left[\eta \mapsto \mathcal{N}_{\varrho_N}^\nu(A)(\eta)^{-1}\right] \in \mathcal{C}^\infty(T^*\partial X \setminus \{0\}, \Psi_{b,c}^{0,0,(\gamma_{\mathrm{lb}},\gamma_{\mathrm{rb}})}(M; {}^{b,c}\Omega^{1/2})),$$

and $I_b\left(\mathcal{N}_{\varrho_N}^\nu(A)(\eta)\right)(\xi + i\mu) \neq 0$ for all $\xi \in \mathbb{R}$ and all $\mu \in [-1/2 - \gamma_{\mathrm{lb}}, 1/2 + \gamma_{\mathrm{rb}}]$. Let $\varepsilon > 0$ be so small that $\gamma_T := \gamma_{\mathrm{lb}} - \varepsilon > -1/2$ and $\gamma_B := \gamma_{\mathrm{rb}} - \varepsilon > -1/2$. Finally, choose γ_F with $n - 1 < \gamma_F < \min\{\gamma_{\mathrm{lb}} + \gamma_{\mathrm{rb}} + n, n\}$

Because of ${}^0\sigma^{(0)}(A)(\zeta) \neq 0$ for all $\zeta \in {}^0S^*X$, there exists $B_0 \in \Psi_0^0(X; {}^0\Omega^{1/2})$ such that we have $R_0 := \mathrm{id} - AB_0 \in \Psi_0^{-\infty}(X; {}^0\Omega^{1/2})$ by Proposition 2.3.7. By Proposition 3.5.3, Proposition 3.6.2, and Proposition 4.4.1, the map
(6.1)
$$\mathcal{N} : T^*\partial X \setminus \{0\} \longrightarrow \Psi_{b,c}^{-\infty,-\infty,(\gamma_{\mathrm{lb}},\gamma_{\mathrm{rb}})}(M; {}^{b,c}\Omega^{1/2}) : \eta \longmapsto \mathcal{N}_{\varrho_N}^\nu(A)(\eta)^{-1}\mathcal{N}_{\varrho_N}^\nu(R_0)(\eta)$$

is smooth; the b-indicial family of $\mathcal{N}(\eta)$ is given by

$$I_b\left(\mathcal{N}(\eta)\right)(z) = I_b\left(\mathcal{N}_{\varrho_N}^\nu(A)(\eta)\right)(z)^{-1} I_b\left(\mathcal{N}_{\varrho_N}^\nu(R_0)(\eta)\right)(z)$$

for $-1/2 - \gamma_{\mathrm{rb}} < \mathrm{Im}\, z < 1/2 + \gamma_{\mathrm{lb}}$; recall that by (4.25) the b-indicial family actually does only depend on $\pi(\eta)$, i.e. it is constant on the fibers of $T^*\partial X \setminus \{0\}$. Using the coordinates (τ, r) near the b-front face ff^b of $M_{b,c}^2$ as explained before (3.11), we can write $\mathcal{N} = \widehat{\mathcal{N}}(\tau, r) \left|d\tau \frac{dr}{r}\right|^{\frac{1}{2}}$ with an appropriate function $\widehat{\mathcal{N}} : (-1,1) \times \mathbb{R}_+ \to \mathbb{C}$.

We are going to show that $(\mathcal{N}, \mathcal{F})$ with $\mathcal{F} : T^*\partial X \to \mathcal{C}^\infty((-1,1))$ given by

$$(6.2) \quad \mathcal{F} : \eta \longmapsto \left[\tau \mapsto \begin{cases} \widehat{\mathcal{N}}(\eta)(\tau, 1) & , \eta \neq \pi(\eta) \\ (1-\tau^2)^{-1/2} \int_{\mathbb{R}_\xi} \left(\frac{1+\tau}{1-\tau}\right)^{i\xi} I_b\left(\mathcal{N}(\eta)\right)(\xi)\, d\xi & , \eta = \pi(\eta) \end{cases}\right]$$

belongs to $\widetilde{\mathfrak{N}}_0^{-\infty,0,(\gamma_T,\gamma_B)}(\delta)$ for any δ with $\gamma_F - (n-1) < \delta \leq \min\{\gamma_{\text{lb}} + \gamma_{\text{rb}} + 1, 1\}$.

By the very definition, $\mathcal{N}$ and $\mathcal{F}$ satisfy (a) of Definition 4.4.7, and it remains to consider $\mathcal{F}$. The smoothness of $\mathcal{F}| : T^*\partial X \setminus \{0\} \to \mathcal{A}^{(\gamma_T,\gamma_B)}([-1,1])$ is a direct consequence of (6.2). Moreover, we have $\mathcal{F}(\pi(\eta)) \in \mathcal{A}^{(\gamma_T,\gamma_B)}([-1,1])$ by Remark 3.5.1 and the choice of γ_T and γ_B.

By Proposition 3.5.3 and Proposition 3.6.2, we can decompose the map $\mathcal{N}$ as $\mathcal{N} = \mathcal{N}_0 + \mathcal{N}_1$ with smooth maps $\mathcal{N}_0 : T^*\partial X \setminus \{0\} \to \widetilde{\Psi}_{b,c}^{-\infty,-\infty,(\gamma_{\text{lb}},\gamma_{\text{rb}})}(M; {}^{b,c}\Omega^{1/2})$ and $\mathcal{N}_1 : T^*\partial X \setminus \{0\} \to \Psi^{-\infty,(\gamma'_{\text{lb}},\gamma'_{\text{rb}})}(M, {}^{b,c}\Omega^{1/2})$. With respect to the coordinates (τ, r) we have $\mathcal{N}_j = \widehat{\mathcal{N}}_j(\tau,r)\left|d\tau\frac{dr}{r}\right|^{\frac{1}{2}}$ with appropriate smooth functions $\widehat{\mathcal{N}}_j$. By the definition of $\Psi^{-\infty,(\gamma'_{\text{lb}},\gamma'_{\text{rb}})}(M, {}^{b,c}\Omega^{1/2})$ there exists a smooth function $H : T^*\partial X \setminus \{0\} \to \mathcal{A}^{(\gamma'_{\text{lb}},\gamma'_{\text{rb}})}(\overline{\mathbb{R}}_+ \times \overline{\mathbb{R}}_+)$ vanishing rapidly as $|(x,x')| \to \infty$ such that $\widehat{\mathcal{N}}_1(\eta)(\tau, r) = H(\eta)(r/2(1+\tau), r/2(1-\tau))$.

Let us first show that $\mathcal{F} : T^*\partial X \to \mathcal{A}^{(\gamma_T,\gamma_B)}([-1,1])$ is continuous at the zero section $\eta = \pi(\eta)$. Since the restriction $\mathcal{F}|_{\partial X}$ to the zero section $\partial X \subseteq T^*\partial X$ is smooth, it suffices to find for each $\varepsilon > 0$, for each compact set $K \subseteq T^*\partial X \setminus \{0\}$ and each continuous seminorm p on the space $\mathcal{A}^{(\gamma_T,\gamma_B)}([-1,1])$ an $r_0 > 0$ such that $p(\mathcal{F}(r\eta) - \mathcal{F}(\pi(\eta))) < \varepsilon$ for all $r < r_0$ and all $\eta \in K$. Indeed, consider the decomposition

$$\mathcal{F}(r\eta) - \mathcal{F}(\pi(\eta)) = \widehat{\mathcal{N}}_0(\eta)(\cdot, r) - \mathcal{F}(\pi(\eta))(\cdot) + \widehat{\mathcal{N}}_1(\eta)(\cdot, r).$$

It remains to note that $\widehat{\mathcal{N}}_0(\eta)$ is smooth up to the b-front-face $\text{ff}^b = \{r = 0\}$ with $\widehat{\mathcal{N}}_0(\eta)|_{\text{ff}^b}(\tau) = \mathcal{F}(\pi(\eta))(\tau)$ by (6.2), and that $r^{-1-\gamma_{\text{lb}}-\gamma_{\text{rb}}}\widehat{\mathcal{N}}_1(\eta)(\cdot, r)$ is bounded in $\mathcal{A}^{(\gamma_T,\gamma_B)}([-1,1])$ uniformly for $\eta \in K$ and $0 \leq r \leq 1$. The identity (4.26) is a direct consequence of (6.2) and the Mellin inversion formula.

Further, (c) of Definition 4.4.7 follows using the decomposition

$$\beta^*\mathcal{F}(r,\eta) = \mathcal{F}(r\eta) = \widehat{\mathcal{N}}_0(\eta)(\cdot, r) + \widehat{\mathcal{N}}_1(\eta)(\cdot, r) \text{ for } r > 0 \text{ and } \eta \in T^*\partial X \setminus \{0\}$$

because $\widehat{\mathcal{N}}_0(\eta)(\cdot, r)$ as well as $\widehat{\mathcal{N}}_1(\eta)(\cdot, r)$ vanish rapidly with values in the space $\mathcal{A}^{(\gamma_T,\gamma_B)}([-1,1])$ as $r \to \infty$.

Finally, for (d) of Definition 4.4.7, we use local coordinates $\vartheta = (y, \widehat{\eta}, \varrho)$ in $\mathcal{P}^*\partial X$ as in (4.29). Thus, for $\{\varrho_0^{-\delta}P(\beta^*\mathcal{F}(\vartheta) - \mathcal{F}(\pi(\vartheta)))\}$ from (4.27) we get

$$\varrho^{-\delta}P(\mathcal{F}(y, \varrho\widehat{\eta})(\tau) - \mathcal{F}(y, 0)(\tau)) =$$
$$\varrho^{-\delta}P\left(\widehat{\mathcal{N}}_0(\widehat{\eta})(\tau,\varrho) - \widehat{\mathcal{N}}_0(\widehat{\eta})(\tau, 0)\right) + \varrho^{-\delta}PH(\widehat{\eta})(\varrho/2(1+\tau), \varrho/2(1-\tau)).$$

The first term on the right hand side is bounded in $\mathcal{A}^{(\gamma_T,\gamma_B)}([-1,1])$ because $\widehat{\mathcal{N}}_0(\widehat{\eta})$ is smooth up to $\varrho = 0$, whereas the second term stays bounded in $\mathcal{A}^{(\gamma_T,\gamma_B)}([-1,1])$ because $H(\widehat{\eta})$ and all its conormal derivatives can be estimated up to a constant by $\varrho^{\gamma_{\text{lb}}+\gamma_{\text{rb}}+1}(1+\tau)^{\gamma_{\text{lb}}+1/2}(1-\tau)^{\gamma_{\text{rb}}+1/2}$, and $\gamma_{\text{lb}} + \gamma_{\text{rb}} + 1 - \delta \geq 0$ by the choice of δ. This finally yields $(\mathcal{N}, \mathcal{F}) \in \widetilde{\mathfrak{N}}_0^{-\infty,0,(\gamma_T,\gamma_B)}(\delta)$ for $\gamma_F - (n-1) < \delta \leq 1$.

Then by Theorem 4.4.8, there exists an operator $B_1 \in \Psi_0^{-\infty,0,\gamma}(X; {}^0\Omega^{1/2})$ with $\mathcal{N}_{\varrho_N}^\nu(B_1)(\eta) = \mathcal{N}(\eta)$ for all $\eta \in T^*\partial X \setminus \{0\}$ where $\gamma = (\gamma_T, \gamma_B, \gamma_F)$ is as above; hence, for $B := B_0 + B_1 \in \Psi_0^{0,0,\gamma}(X; {}^0\Omega^{1/2})$, we have

$$R_1 := \text{id} - AB = R_0 - AB_1 \in \Psi_0^{-\infty,0,\gamma}(X; {}^0\Omega^{1/2})$$

with $\mathcal{N}_{\varrho_N}^\nu(R_1)(\eta) = \mathcal{N}_{\varrho_N}^\nu(R_0)(\eta) - \mathcal{N}_{\varrho_N}^\nu(A)(\eta)\mathcal{N}(\eta) = 0$ for all $\eta \in T^*\partial X \setminus \{0\}$, i.e. we have in fact $R_1 \in \Psi_0^{-\infty,-1,\gamma}(X; {}^0\Omega^{1/2})$ and $R_1 : L^2(X, {}^0\Omega^{1/2}) \to L^2(X, {}^0\Omega^{1/2})$

is compact by Theorem 5.2.2. Similarly, we obtain also a left-inverse of A up to compact remainders, and $A : L^2(X, {}^0\Omega^{1/2}) \to L^2(X, {}^0\Omega^{1/2})$ is Fredholm. $\square$

By Theorem 6.2.1, the algebra $\Psi_0^0(X; {}^0\Omega^{1/2})$ of zeroth order 0-operators is an algebra with a $\mathcal{K}(L^2(X, {}^0\Omega^{1/2}))$-symbolic structure in the C^*-algebra $\mathcal{L}(L^2(X, {}^0\Omega^{1/2}))$ in the sense of [40, Section 2.4], with the symbol morphism given by the homogeneous principal symbol ${}^0\sigma^{(0)}$ and the reduced normal operator $\mathcal{N}_{\varrho_N}^\nu$. We thus obtain a characterization of the compact operators in $\Psi_0^0(X; {}^0\Omega^{1/2})$ as well [40, Proposition 2.30].

COROLLARY 6.2.2. *Let $A \in \Psi_0^0(X; {}^0\Omega^{1/2})$ be arbitrary. Then the bounded operator $A : L^2(X, {}^0\Omega^{1/2}) \to L^2(X, {}^0\Omega^{1/2})$ is compact if and only if ${}^0\sigma^{(0)}(A) = 0$ and $\mathcal{N}_{\varrho_N}^\nu(A)(\eta) = 0$ for all $\eta \in T^*\partial X \setminus \{0\}$.*

Using Proposition 4.5.3 we easily extend these results to weighted spaces.

COROLLARY 6.2.3. *Let $\mathfrak{a} \in \mathbb{R}$ and $A \in \Psi_0^0(X; {}^0\Omega^{1/2})$ be arbitrary. Then we have*

(a) *The operator $A : \varrho_N^{\mathfrak{a}} L^2(X, {}^0\Omega^{1/2}) \to \varrho_N^{\mathfrak{a}} L^2(X, {}^0\Omega^{1/2})$ is Fredholm if and only if we have ${}^0\sigma^{(0)}(A)(\zeta) \neq 0$ for all $\zeta \in {}^0S^*X$ and the reduced normal operator $\mathcal{N}_{\varrho_N}^\nu(A)(\eta) : \varrho_0^{\mathfrak{a}} L^2(M, {}^{b,c}\Omega^{1/2}) \to \varrho_0^{\mathfrak{a}} L^2(M, {}^{b,c}\Omega^{1/2})$ is invertible for all $\eta \in T^*\partial X \setminus \{0\}$.*

(b) *The operator $A : \varrho_N^{\mathfrak{a}} L^2(X, {}^0\Omega^{1/2}) \to \varrho_N^{\mathfrak{a}} L^2(X, {}^0\Omega^{1/2})$ is compact if and only if we have ${}^0\sigma^{(0)}(A)(\zeta) = 0$ for all $\zeta \in {}^0S^*X$ and $\mathcal{N}_{\varrho_N}^\nu(A)(\eta) = 0$ for all $\eta \in T^*\partial X \setminus \{0\}$.*

6.3. Characterization of Fredholm operators in $\Psi_0^{m,k}(X; {}^0\Omega^{1/2})$

PROPOSITION 6.3.1. *Let $\mathfrak{a} \in \mathbb{R}$ and $A \in \Psi_0^m(X; {}^0\Omega^{1/2})$ be with ${}^0\sigma^{(m)}(A)(\zeta) \neq 0$ for all $\zeta \in {}^0S^*X$. Then the operator $A : \varrho_N^{\mathfrak{a}} H_0^s(X, {}^0\Omega^{1/2}) \longrightarrow \varrho_N^{\mathfrak{a}} H_0^{s-m}(X, {}^0\Omega^{1/2})$ is Fredholm provided the reduced normal operator*

$$\mathcal{N}_{\varrho_N}^\nu(A)(\eta) : \varrho_0^{\mathfrak{a}} \varrho_1^{\mathfrak{a}_1} H_{b,c}^m(M, {}^{b,c}\Omega^{1/2}) \to \varrho_0^{\mathfrak{a}} \varrho_1^{\mathfrak{a}_1 - m} L^2(M, {}^{b,c}\Omega^{1/2})$$

is invertible for all $\eta \in T^*\partial X \setminus \{0\}$ and some, hence any $\mathfrak{a}_1 \in \mathbb{R}$.

Proof: By conjugating with $\varrho_N^{\mathfrak{a}}$ we can assume $\mathfrak{a} = 0$. Then the proof continues exactly as the proof of Theorem 6.2.1. $\square$

PROPOSITION 6.3.2. *For $k, m \in \mathbb{R}$, there exist $\Lambda_{m,k} \in \Psi_0^{m,k}(X; {}^0\Omega^{1/2})$ such that $\Lambda_{m,k} = (\Lambda_{m,k})^\sharp$ and the operator $\Lambda_{m,k} : \dot{\mathcal{C}}^\infty(X, {}^0\Omega^{1/2}) \to \dot{\mathcal{C}}^\infty(X, {}^0\Omega^{1/2})$ extends to an isomorphism*

$$\Lambda_{m,k}^{s,\mathfrak{a}} = \Lambda_{m,k} : \varrho_N^{\mathfrak{a}} H_0^s(X, {}^0\Omega^{1/2}) \longrightarrow \varrho_N^{\mathfrak{a}-k} H_0^{s-m}(X, {}^0\Omega^{1/2})$$

for any $\mathfrak{a}, s \in \mathbb{R}$. Moreover,

$$\mathcal{N}_{\varrho_N}^\nu(\varrho_N^k \Lambda_{m,k})(\eta) : \varrho_0^{\mathfrak{a}} \varrho_1^s H_{b,c}^s(M, {}^{b,c}\Omega^{1/2}) \to \varrho_0^{\mathfrak{a}} \varrho_1^{s-m} H_{b,c}^{s-m}(M, {}^{b,c}\Omega^{1/2})$$

is invertible for all $\mathfrak{a}, s \in \mathbb{R}$ and all $\eta \in T^*\partial X \setminus \{0\}$, and we have ${}^0\sigma^{(m,k)}(\Lambda_{m,k}) = 1$.

Proof: Since $\Lambda_{m,k} := \varrho_N^{-k} \Lambda_{m,0}$ has the desired properties provided $\Lambda_{m,0}$ is as in the proposition, we can assume $k = 0$. By the exactness of (2.4), we find $B \in \Psi_0^{m/2}(X; {}^0\Omega^{1/2})$ with ${}^0\sigma^{(m/2)}(B) = 1$. By Proposition 2.3.7, there exists $B_1 \in \Psi_0^{-m/2}(X; {}^0\Omega^{1/2})$ such that $R := \mathrm{id} - B_1 B \in \Psi_0^{-\infty}(X; {}^0\Omega^{1/2})$. Then

$$\Lambda_{m,0} := A := B^\sharp B + R^\sharp R \in \Psi_0^m(X; {}^0\Omega^{1/2})$$

has the desired properties. Indeed, first note that we have $^0\sigma^{(m)}(\Lambda_{m,0}) = 1$, thus, $^{b,c}\sigma^{(m,m)}(\mathcal{N}_{\varrho_N}^\nu(A)(\eta)) = 1$ and $I_c^{(m)}(\mathcal{N}_{\varrho_N}^\nu(A)(\eta)) = 1$ by Proposition 4.4.4. Moreover, $I_b\left(\mathcal{N}_{\varrho_N}^\nu(A)(\eta)\right)(\xi) \neq 0$ for all $\xi \in \mathbb{R}$ by (4.21) and the choice of R, therefore, $\mathcal{N}_{\varrho_N}^\nu(A)(\eta) : \varrho_1^s H_{b,c}^s(M, {}^{b,c}\Omega^{1/2}) \to \varrho_1^{s-m} H_{b,c}^{s-m}(M, {}^{b,c}\Omega^{1/2})$ is Fredholm for any $\eta \in T^*\partial X \setminus \{0\}$. On the other hand, if $f \in H_{b,c}^s(M, {}^{b,c}\Omega^{1/2})$ is in the kernel of $\mathcal{N}_{\varrho_N}^\nu(A)(\eta)$, we have $\mathcal{N}_{\varrho_N}^\nu(B)(\eta)f = \mathcal{N}_{\varrho_N}^\nu(R)(\eta)f = 0$ by Proposition 4.5.1 and Proposition 4.5.2, thus, $f = 0$ by the choice of R and B.

Since the kernel of the Hilbert space adjoint
$$(\mathcal{N}_{\varrho_N}^\nu(A)(\eta))^* : \varrho_1^{s-m} H_{b,c}^{s-m}(M, {}^{b,c}\Omega^{1/2}) \to \varrho_1^s H_{b,c}^s(M, {}^{b,c}\Omega^{1/2})$$
is in fact isomorphic to $N(\mathcal{N}_{\varrho_N}^\nu(A)(\eta)^\sharp) = N(\mathcal{N}_{\varrho_N}^\nu(A)(\eta))$, $\mathcal{N}_{\varrho_N}^\nu(A)(\eta)$ is an isomorphism. By Proposition 6.3.1, $A = \Lambda_{m,0} : H_0^s(X, {}^0\Omega^{1/2}) \to H_0^{s-m}(X, {}^0\Omega^{1/2})$ is a Fredholm operator. Exactly as for $\mathcal{N}_{\varrho_N}^\nu(A)(\eta)$ we show that it is actually an isomorphism which completes the proof. □

These *order-reducing operators* can be used to reduce the characterization of the Fredholm operators in the general case $\Psi_0^{m,k}(X; {}^0\Omega^{1/2})$ to the case $\Psi_0^0(X; {}^0\Omega^{1/2})$ and we get:

THEOREM 6.3.3. *Let $\mathfrak{a}, s \in \mathbb{R}$ and $A \in \Psi_0^{m,k}(X; {}^0\Omega^{1/2})$ be arbitrary.*

(a) *The operator $A : \varrho_N^\mathfrak{a} H_0^s(X, {}^0\Omega^{1/2}) \to \varrho_N^{\mathfrak{a}-k} H_0^{s-m}(X, {}^0\Omega^{1/2})$ is Fredholm if and only if $^0\sigma^{(m,k)}(A)(\zeta) \neq 0$ for all $\zeta \in {}^0S^*X$ and the reduced normal operator*

(6.3) $$\mathcal{N}_{\varrho_N}^\nu(\varrho_N^k A)(\eta) : \varrho_0^\mathfrak{a} \varrho_1^s H_{b,c}^s(M, {}^{b,c}\Omega^{1/2}) \to \varrho_0^\mathfrak{a} \varrho_1^{s-m} H_{b,c}^{s-m}(M, {}^{b,c}\Omega^{1/2})$$

is invertible for all $\eta \in T^*\partial X \setminus \{0\}$.

(b) *The operator $A : \varrho_N^\mathfrak{a} H_0^s(X, {}^0\Omega^{1/2}) \to \varrho_N^{\mathfrak{a}-k} H_0^{s-m}(X, {}^0\Omega^{1/2})$ is compact if and only if $^0\sigma^{(m,k)}(A)(\zeta) = 0$ for all $\zeta \in {}^0S^*X$ and $\mathcal{N}_{\varrho_N}^\nu(\varrho_N^k A)(\eta) = 0$ for all $\eta \in T^*\partial X \setminus \{0\}$.*

Note that by (4.33) we can replace $\mathcal{N}_{\varrho_N}^\nu(\varrho_N^k A)(\eta)$ by $\mathcal{N}_{\varrho_N}^\nu(A\varrho_N^k)(\eta)$ if we shift the weight in (6.3) accordingly.

Proof: We simply apply Theorem 6.2.1 to $B := \Lambda_{s-m,\mathfrak{a}-k} A \Lambda_{-s,-\mathfrak{a}} \in \Psi_0^0(X; {}^0\Omega^{1/2})$ where $\Lambda_{m,k} \in \Psi_0^{m,k}(X; {}^0\Omega^{1/2})$ is as in Proposition 6.3.2. For the invertibility of the reduced normal operator we use the identity

$$\mathcal{N}_{\varrho_N}^\nu(B)(\eta)$$
$$= \mathcal{N}_{\varrho_N}^\nu\left(\varrho_N^{k-\mathfrak{a}}\left(\varrho_N^{\mathfrak{a}-k}\Lambda_{s-m,\mathfrak{a}-k}\right)\varrho_N^{\mathfrak{a}-k}\left(\varrho_N^{-\mathfrak{a}}(\varrho_N^k A)\varrho_N^\mathfrak{a}\right)\left(\varrho_N^{-\mathfrak{a}},\Lambda_{s,-\mathfrak{a}}\right)\right)(\eta)$$
$$= \varrho_0^{k-\mathfrak{a}}\mathcal{N}_{\varrho_N}^\nu\left(\varrho_N^{\mathfrak{a}-k}\Lambda_{s-m,\mathfrak{a}-k}\right)(\eta)\varrho_0^{-k}\mathcal{N}_{\varrho_N}^\nu\left(\varrho_N^k A\right)(\eta)\varrho_0^\mathfrak{a}\mathcal{N}_{\varrho_N}^\nu\left(\varrho_N^{-\mathfrak{a}}\Lambda_{-s,-\mathfrak{a}}\right)(\eta).$$
□

6.4. General bundles

For the sake of completeness, let us note that similar characterizations of the Fredholm property hold also for the case of 0-operators acting between section of vector bundles. Again, we leave the details to the reader.

Part 2

Algebras of 0-pseudodifferential operators of order 0

CHAPTER 7

C^*-algebras of 0-pseudodifferential operators

Throughout this chapter let $\mathfrak{a} \in \mathbb{R}$ be a fixed weight.

7.1. Solvable C^*-algebras

Recall that a C^*-algebra $\mathcal{B}$ is said to be *solvable* if there exists a finite sequence

(7.1) $$\mathcal{B} = \mathcal{J}_{\ell+1} \supseteq \mathcal{J}_\ell \supseteq \cdots \supseteq \mathcal{J}_1 \supseteq \mathcal{J}_0 = \{0\}$$

of closed ideals such that $\mathcal{J}_{k+1}/\mathcal{J}_k \cong \mathcal{C}_0(T_k, \mathcal{K}(H_k))$ for some locally compact Hausdorff space T_k and some separable Hilbert space H_k [17]. Moreover, the *composition series* (7.1) is said to be *solving of length ℓ*, and the smallest length of a solving composition series is, by definition, the *length of $\mathcal{B}$*. A solvable C^*-algebra is of type I, in particular it is nuclear. For a brief review on solvable operator algebras in analysis we recommend [83].

The *spectrum* $\widehat{\mathcal{B}}$ of a C^*-algebra $\mathcal{B}$ is, by definition, the set of all unitary equivalence classes of non-zero irreducible representations of $\mathcal{B}$; it is equipped with a canonical topology, the *Jacobson topology* $\mathcal{T}_{\mathcal{J}}(\mathcal{B})$. Note that the spectrum of a solvable C^*-algebra with a solving composition series (7.1) is given as a set by

(7.2) $$\widehat{\mathcal{B}} = \biguplus_{k=0}^{\ell} T_k,$$

with irreducible representations induced by point evaluations. A description of the Jacobson topology, however, requires some more analysis. For more details on the spectrum of C^*-algebras we refer to the classical monograph [15].

7.2. The reduced normal operator on $S^*\partial X$

By Corollary 4.3.1, the reduced normal operator $\mathcal{N}_{\varrho_N}^\nu(A)$ of a 0-pseudodifferential operator $A \in \Psi_0^0(X; {}^0\Omega^{1/2})$ is a smooth function

$$\mathcal{N}_{\varrho_N}^\nu(A) : T^*\partial X \setminus \{0\} \to \mathcal{L}(\varrho_0^{\mathfrak{a}} L^2(M, {}^{b,c}\Omega^{1/2}))$$

satisfying $\mathcal{N}_{\varrho_N}^\nu(A)(t\eta) = Q_t \mathcal{N}_{\varrho_N}^\nu(A)(\eta) Q_{t^{-1}}$ for all $t > 0$; thus, there exists a smooth vector bundle $\mathcal{E}_0 \to S^*\partial X$ such that the reduced normal operator $\mathcal{N}_{\varrho_N}^\nu(A)$ induces a continuous section $\mathcal{N}_{\varrho_N}^\nu(A) : S^*\partial X \to \mathcal{E}_0$ again denoted by $\mathcal{N}_{\varrho_N}^\nu(A)$ for simplicity. The vector bundle $\mathcal{E}_0 \to S^*\partial X$ is actually given as the set of all pairs

$$(\eta, a) \in T^*\partial X \setminus \{0\} \times \mathcal{L}(\varrho_0^{\mathfrak{a}} L^2(M, {}^{b,c}\Omega^{1/2}))$$

modulo the equivalence relation

$$(\eta, a) \sim (\eta', a') :\Longleftrightarrow \eta' = t\eta \text{ and } a' = Q_t a Q_{t^{-1}} \text{ for some } t > 0.$$

Let $s : S^*\partial X \to T^*\partial X \setminus \{0\}$ be a smooth embedding that is additionally a right inverse of the natural projection $\widehat{\cdot} : T^*\partial X \setminus \{0\} \to S^*\partial X : \eta \mapsto \widehat{\eta}$ such that

$$F_s : S^*\partial X \times \mathbb{R}_+ \longrightarrow T^*\partial X \setminus \{0\} : (\widehat{\eta}, t) \longmapsto ts(\widehat{\eta})$$

is a diffeomorphism. This is equivalent to the embedding s being transversal to the radial direction in every point. Then s yields a trivialization

$$\mathcal{E}_0 \xrightarrow{\cong} S^*\partial X \times \mathcal{L}(\varrho_0^{\mathfrak{a}} L^2(M, {}^{b,c}\Omega^{1/2})).$$

Note that such a section s can for instance be given by the choice of a Riemannian metric on $T^*\partial X$. However, to avoid unnecessary notational complications we fix such a section s, and think of the reduced normal operator as a smooth function $\mathcal{N}^\nu_{\varrho_N}(A) : S^*\partial X \to \mathcal{L}(\varrho_0^{\mathfrak{a}} L^2(M, {}^{b,c}\Omega^{1/2})) : \widehat{\eta} \mapsto \mathcal{N}^\nu_{\varrho_N}(A)(s(\widehat{\eta}))$. Moreover, s determines a smooth map $|\cdot|_s : T^*\partial X \setminus \{0\} \to \overline{\mathbb{R}}_+$ satisfying $\eta = |\eta|_s s(\widehat{\eta})$ for all $\eta \in T^*\partial X \setminus \{0\}$ simply by projecting the inverse of F_s onto the second factor. With this notation, we obtain for $\eta \in T^*\partial X \setminus \{0\}$

(7.3) $$\mathcal{N}^\nu_{\varrho_N}(A)(\eta) = \mathcal{N}^\nu_{\varrho_N}(A)(|\eta|_s s(\widehat{\eta})) = Q_{|\eta|_s} \mathcal{N}^\nu_{\varrho_N}(A)(s(\widehat{\eta})) Q_{|\eta|_s^{-1}}.$$

For the sake of notational simplicity, we will usually omit the embedding s.

7.3. Extension of the symbolic structure

By Lemma 5.1.4, $\Psi_0^0(X; {}^0\Omega^{1/2})$ is a symmetric subalgebra of the C^*-algebra $\mathcal{L}(\varrho_N^{\mathfrak{a}} L^2(X, {}^0\Omega^{1/2}))$, by Corollary 6.2.3, it is an algebra with a $\mathcal{K}(\varrho_N^{\mathfrak{a}} L^2(X, {}^0\Omega^{1/2}))$-symbolic structure in the sense of [40]; the symbol morphism is given by

$$\tau_0 := ({}^0\sigma^{(0)}, \mathcal{N}^\nu_{\varrho_N}) : \Psi_0^0(X; {}^0\Omega^{1/2}) \longrightarrow \mathcal{C}({}^0S^*X) \oplus \mathcal{C}(S^*\partial X, \mathcal{B}^{(\mathfrak{a})}_{b,c}(M, {}^{b,c}\Omega^{1/2})) =: Q^{(\mathfrak{a})}$$

with $\mathcal{B}^{(\mathfrak{a})}_{b,c}(M, {}^{b,c}\Omega^{1/2}) := \mathcal{B}^{(\mathfrak{a},0)}_{b,c}(M, {}^{b,c}\Omega^{1/2})$. Let $\mathcal{B}^{(\mathfrak{a})}_0(X, {}^0\Omega^{1/2})$ be the closure of $\Psi_0^0(X; {}^0\Omega^{1/2})$ in the C^*-algebra $\mathcal{L}(\varrho_N^{\mathfrak{a}} L^2(X, {}^0\Omega^{1/2}))$. Note that the C^*-structure of the algebra $\mathcal{B}^{(\mathfrak{a})}_0(X, {}^0\Omega^{1/2})$ depends on the choice of the defining function ϱ_N as long as $\mathfrak{a} \neq 0$. Since operators in

$$\Psi_0^{-\infty,-\infty}(X; {}^0\Omega^{1/2}) := \bigcap_{m \in \mathbb{R}} \bigcap_{k \in \mathbb{R}} \Psi_0^{m,k}(X; {}^0\Omega^{1/2})$$

are exactly those with Schwartz kernels in the space $\dot{\mathcal{C}}^\infty(X^2, {}^0\Omega^{1/2} \boxtimes {}^0\Omega^{1/2})$, we have $\mathcal{K}(\varrho_N^{\mathfrak{a}} L^2(X, {}^0\Omega^{1/2})) \subseteq \mathcal{B}^{(\mathfrak{a})}_0(X, {}^0\Omega^{1/2})$, and [40, Proposition 2.31] shows that the symbol morphism τ_0 extends continuously to $\mathcal{B}^{(\mathfrak{a})}_0(X, {}^0\Omega^{1/2})$ leading to the following short exact sequence of C^*-algebras.

(7.4) $$0 \longrightarrow \mathcal{K}(\varrho_N^{\mathfrak{a}} L^2(X, {}^0\Omega^{1/2})) \longrightarrow \mathcal{B}^{(\mathfrak{a})}_0(X, {}^0\Omega^{1/2}) \xrightarrow{\tau_0} Q_0^{(\mathfrak{a})} \longrightarrow 0,$$

where the C^*-algebra $Q_0^{(\mathfrak{a})}$ is given as the norm closure of $\tau_0(\Psi_0^0(X; {}^0\Omega^{1/2}))$ in the C^*-algebra $Q^{(\mathfrak{a})}$. Let us denote the composition of τ_0 with the projection onto the first resp. the second component of $Q^{(\mathfrak{a})}$ by ${}^0\sigma^{(0)}$ resp. $\mathcal{N}^{\nu,\mathfrak{a}}_{\varrho_N}$.

REMARK 7.3.1. Note that ${}^0\sigma^{(0)} : \mathcal{B}^{(\mathfrak{a})}_0(X, {}^0\Omega^{1/2}) \to \mathcal{C}({}^0S^*X)$ is onto because its range $R({}^0\sigma^{(0)})$ contains the dense subset ${}^0\sigma^{(0)}(\Psi_0^0(X; {}^0\Omega^{1/2})) = \mathcal{C}^\infty({}^0S^*X)$ by (2.4), and is closed

The next result is a first step towards understanding the algebra $Q_0^{(\mathfrak{a})}$ of *joint 0-symbols*.

PROPOSITION 7.3.2. *The C^*-algebra $Q_0^{(\mathfrak{a})}$ is contained in the C^*-algebra of all pairs $(f, \mathcal{N}) \in Q^{(\mathfrak{a})}$ satisfying for each $\eta \in T^*\partial X \setminus \{0\}$ the following compatibility conditions:*

$$(7.5) \quad {}^{b,c}\sigma^{(0,0)}(\mathcal{N}(\widehat{\eta}))(z,\xi) = f_0\left(j_{\varrho_N}(0,(\pi(\eta),\xi))\right) \text{ for } (z,\xi) \in {}^{b,c}T^*M \setminus \{0\}$$

$$(7.6) \quad I_c(\mathcal{N}(\widehat{\eta}))(\xi/|\eta|_s) = f_0\left(j_{\varrho_N}(\eta,(\pi(\eta),\xi))\right) \text{ for } \xi \in \mathbb{R}$$

$$(7.7) \quad I_b^{(\mathfrak{a})}(\mathcal{N}(\widehat{\eta}_1)) = I_b^{(\mathfrak{a})}(\mathcal{N}(\widehat{\eta}_2)) \text{ whenever } \pi(\widehat{\eta}_1) = \pi(\widehat{\eta}_2).$$

Here, $f_0 \in \mathcal{C}({}^0T^*X \setminus \{0\})$ is the function, homogeneous of degree 0 in the fibers that corresponds naturally to $f \in \mathcal{C}({}^0S^*X)$, i.e. $f_0(\zeta) = f(\widehat{\zeta})$ for all $\zeta \in {}^0T^*X \setminus \{0\}$.

Proof: By Proposition 4.4.4, for $A \in \Psi_0^0(X; {}^0\Omega^{1/2})$, the pair $({}^0\sigma^{(0)}(A), \mathcal{N}_{\varrho_N}^\nu(A))$ satisfies the compatibility conditions (7.5) and (7.6). On the other hand, the maps

$$\begin{aligned}
{}^{b,c}\sigma^{(0,0)} \circ \mathcal{N}_{\varrho_N}^{\nu,\mathfrak{a}}(\cdot)(\widehat{\eta}) : \mathcal{B}_0^{(\mathfrak{a})}(X, {}^0\Omega^{1/2}) &\longrightarrow \mathcal{C}({}^{b,c}S^*M), \\
I_c(\cdot) \circ \mathcal{N}_{\varrho_N}^{\nu,\mathfrak{a}}(\cdot)(\widehat{\eta}) : \mathcal{B}_0^{(\mathfrak{a})}(X, {}^0\Omega^{1/2}) &\longrightarrow \mathcal{B}(;\mathbb{R}), \\
I_b^{(\mathfrak{a})}(\cdot) \circ \mathcal{N}_{\varrho_N}^{\nu,\mathfrak{a}}(\cdot)(\widehat{\eta}) : \mathcal{B}_0^{(\mathfrak{a})}(X, {}^0\Omega^{1/2}) &\longrightarrow \mathcal{B}(;\Gamma_\mathfrak{a}) \text{ and} \\
{}^0\sigma^{(0)} : \mathcal{B}_0^{(\mathfrak{a})}(X, {}^0\Omega^{1/2}) &\longrightarrow \mathcal{C}({}^0S^*X)
\end{aligned}$$

are continuous. This completes the proof. $\square$

REMARK 7.3.3. (a) Note that in particular the homogeneous principal symbol of $\mathcal{N}(\widehat{\eta})$ firstly does not depend on $z \in M$, and secondly does depend only on the projection $\pi(\eta) \in \partial X$ of $\eta \in T^*\partial X \setminus \{0\}$.

(b) By Proposition 3.7.1, it suffices to require conditions (7.5) and (7.6) for all $\widehat{\eta} \in S^*\partial X = s(S^*\partial X) \subseteq T^*\partial X \setminus \{0\}$ only.

A precise description of $Q_0^{(\mathfrak{a})}$ is given in Proposition 7.5.1 below. First we need a better understanding of the reduced normal operator.

7.4. The C^*-algebra generated by the reduced normal operator

Let us denote by

$$\mathcal{B}_{b,c}^{(\mathfrak{a})}(M, {}^{b,c}\Omega^{1/2}; S^*\partial X) := R\left(\mathcal{N}_{\varrho_N}^{\nu,\mathfrak{a}} : \mathcal{B}_0^{(\mathfrak{a})}(X, {}^0\Omega^{1/2}) \to \mathcal{C}(S^*\partial X, \mathcal{B}_{b,c}^{(\mathfrak{a})}(M, {}^{b,c}\Omega^{1/2}))\right)$$

the range of the reduced normal operator. Since the range of a $*$-morphism between C^*-algebras is always closed, $\mathcal{B}_{b,c}^{(\mathfrak{a})}(M, {}^{b,c}\Omega^{1/2}; S^*\partial X)$ is a C^*-subalgebra of the C^*-algebra $\mathcal{C}(S^*\partial X, \mathcal{B}_{b,c}^{(\mathfrak{a})}(M, {}^{b,c}\Omega^{1/2}))$.

LEMMA 7.4.1. *Let $\mathcal{N} \in \mathcal{B}_{b,c}^{(\mathfrak{a})}(M, {}^{b,c}\Omega^{1/2}; S^*\partial X)$ be arbitrary.*

(a) *The homogeneous principal symbol ${}^{b,c}\sigma^{(0,0)}(\mathcal{N}(\widehat{\eta}))(z,\xi)$ depends only on $\pi(\widehat{\eta}) \in \partial X$ and the sign of $\xi \in \mathbb{R} \setminus \{0\}$, i.e. it does neither depend on $z \in M$ nor on the point in the fiber over $\pi(\widehat{\eta})$.*

(b) *For all $\eta \in T^*\partial X \setminus \{0\}$, we have*

$$(7.8) \quad \sigma_{\mathcal{B}}^{(0)}(I_c(\mathcal{N}(\widehat{\eta})))(\pm 1) = {}^{b,c}\sigma^{(0,0)}(\mathcal{N}(\widehat{\eta}))(1, \pm 1) = \\
= {}^{b,c}\sigma^{(0,0)}(\mathcal{N}(\widehat{\eta}))(0, \pm 1) = \sigma_{\mathcal{B}}^{(0)}(I_b^{(\mathfrak{a})}(\mathcal{N}(\widehat{\eta})))(\pm 1).$$

(c) *The function $I_b^{(\mathfrak{a})}(\mathcal{N}(\widehat{\eta})) \in \mathcal{B}(;\Gamma_\mathfrak{a})$ depends only on $\pi(\widehat{\eta}) \in \partial X$.*

Proof: (a) follows immediately from Proposition 7.3.2, (b) from (3.15), (3.16) and Proposition 7.3.2, and (c), finally, is again a consequence of Proposition 7.3.2. □

LEMMA 7.4.2.   (a) $J_1 := \mathcal{C}(S^*\partial X, \mathcal{K}(\varrho_0^{\mathfrak{a}} L^2(M, {}^{b,c}\Omega^{1/2})))$ is a closed, two-sided ideal in $\mathcal{B}_{b,c}^{(\mathfrak{a})}(M, {}^{b,c}\Omega^{1/2}; S^*\partial X)$. Moreover, we have
$$J_1 \subseteq \mathcal{N}_{\varrho_N}^{\nu}\left(N({}^0\sigma^{(0)} : \mathcal{B}_0^{(\mathfrak{a})}(X, {}^0\Omega^{1/2}) \to \mathcal{C}({}^0S^*X))\right),$$
i.e. for each $\mathcal{N} \in J_1$, we find $A \in N({}^0\sigma^{(0)})$ with $\mathcal{N}_{\varrho_N}^{\nu}(A) = \mathcal{N}$.

(b) The $*$-morphism $A \longmapsto \left[(\widehat{\eta}, \xi) \mapsto I_c\left(\mathcal{N}_{\varrho_N}^{\nu}(A)(\widehat{\eta})\right)(\xi)\right]$ induces a map
$$I_c \circ \mathcal{N}_{\varrho_N}^{\nu,\mathfrak{a}} : N({}^{b,c}\sigma^{(0,0)} \circ \mathcal{N}_{\varrho_N}^{\nu,\mathfrak{a}}) \cap N(I_b^{(\mathfrak{a})} \circ \mathcal{N}_{\varrho_N}^{\nu,\mathfrak{a}}) \longrightarrow \mathcal{C}_0(S^*\partial X \times \mathbb{R}_\xi)$$
that is well-defined and onto.

(c) The $*$-morphism $A \longmapsto \left[(q, \xi) \mapsto I_b^{(\mathfrak{a})}\left(\mathcal{N}_{\varrho_N}^{\nu,\mathfrak{a}}(A)(\widehat{\eta})\right)(\xi)\right]$ for any $\widehat{\eta} \in S^*\partial X$ with $\pi(\widehat{\eta}) = q \in \partial X$ induces a map
$$I_b^{(\mathfrak{a})} \circ \mathcal{N}_{\varrho_N}^{\nu,\mathfrak{a}} : N({}^0\sigma^{(0)}) \longrightarrow \mathcal{C}_0(\partial X \times \Gamma_\mathfrak{a})$$
that is well-defined and onto. Moreover, we have

(7.9) $N({}^0\sigma^{(0)} : \mathcal{B}_0^{(\mathfrak{a})}(X, {}^0\Omega^{1/2}) \to \mathcal{C}({}^0S^*X)) \subseteq N({}^{b,c}\sigma^{(0,0)} \circ \mathcal{N}_{\varrho_N}^{\nu,\mathfrak{a}}) \cap N(I_c \circ \mathcal{N}_{\varrho_N}^{\nu,\mathfrak{a}})$.

(d) Let us denote by J_2 the set of all $\mathcal{N} \in \mathcal{C}(S^*\partial X, \mathcal{B}_{b,c}^{(\mathfrak{a})}(M, {}^{b,c}\Omega^{1/2}))$ satisfying ${}^{b,c}\sigma^{(0,0)}(\mathcal{N}(\widehat{\eta})) = 0$ for all $\widehat{\eta} \in S^*\partial X$, and (7.7). Then J_2 is a closed, two-sided ideal in $\mathcal{B}_{b,c}^{(\mathfrak{a})}(M, {}^{b,c}\Omega^{1/2}; S^*\partial X)$.

Proof: Note that for the surjectivity of a morphism between C^*-algebras it suffices to find a dense subset in its range because the range of a $*$-morphism is always closed.

(a) Let E_1 be the space $\mathcal{C}^\infty(S^*\partial X)\widehat{\otimes}_\pi \dot{\mathcal{C}}^\infty([-1,1], \Omega^{1/2})\widehat{\otimes}_\pi \mathcal{C}_c^\infty(\mathbb{R}_+, {}^b\Omega^{1/2})$ and $E_2 := \mathcal{C}^\infty(S^*\partial X, \mathcal{K}(\varrho_0^{\mathfrak{a}} L^2(M, {}^{b,c}\Omega^{1/2})))$. Then we have continuous and dense embeddings $E_1 \hookrightarrow E_2 \hookrightarrow J_1$. Let $\mathcal{N} = \widehat{\mathcal{N}}(\widehat{\eta}; \tau, r)\left|d\tau\frac{dr}{r}\right|^{\frac{1}{2}} \in E_1$ be arbitrary, and define

$$\mathcal{F}: T^*\partial X \times [-1,1] \longrightarrow \mathbb{C} : (\eta, \tau) \longmapsto \begin{cases} \widehat{\mathcal{N}}(s(\widehat{\eta}); \tau, |\eta|_s) & , \eta \neq \pi(\eta) \in \partial X \\ 0 & , \eta = \pi(\eta) \in \partial X \end{cases}.$$

Then we have $\mathcal{F} \in \mathscr{S}(T^*\partial X)\widehat{\otimes}_\pi \dot{\mathcal{C}}^\infty([-1,1])$, and the extension of $\mathcal{N}$ to $T^*\partial X \setminus \{0\}$ via (7.3) belongs to $\mathcal{N}_{\varrho_N}^{\nu}(\Psi_0^{-\infty}(X; {}^0\Omega^{1/2}))$ by Proposition 4.4.1, i.e. $J_1 \subseteq \mathcal{B}_{b,c}^{(\mathfrak{a})}(M, {}^{b,c}\Omega^{1/2}; S^*\partial X)$.

(b) By Lemma 7.4.1, for $A \in \mathcal{B}_0^{(\mathfrak{a})}(X, {}^0\Omega^{1/2})$, we have
$$\sigma_\mathcal{B}^{(0)}\left(I_c\left(\mathcal{N}_{\varrho_N}^{\nu,\mathfrak{a}}(A)(\widehat{\eta})\right)\right)(\pm 1) = {}^{b,c}\sigma^{(0,0)}(\mathcal{N}_{\varrho_N}^{\nu,\mathfrak{a}}(A)(\widehat{\eta}))(1, \pm 1) = 0,$$
hence we obtain $I_c\left(\mathcal{N}_{\varrho_N}^{\nu,\mathfrak{a}}(A)(\widehat{\eta})\right) \in \mathcal{C}_0(\mathbb{R})$, and the map $I_c \circ \mathcal{N}_{\varrho_N}^{\nu,\mathfrak{a}}$ is well-defined. We are now going to show that the range of $I_c \circ \mathcal{N}_{\varrho_N}^{\nu,\mathfrak{a}}$ contains $\mathcal{C}^\infty(S^*\partial X) \otimes \mathcal{C}_c^\infty(\mathbb{R}_\xi)$ which is dense in $\mathcal{C}_0(S^*\partial X \times \mathbb{R}_\xi)$. By a partition of unity, it suffices to find for any $h_1 \otimes h_2 \in \mathcal{C}_c^\infty(\mathbb{R}_{y'}^{n-1} \times S_{\widehat{\eta}}^{n-2}) \otimes \mathcal{C}_c^\infty(\mathbb{R}_\xi)$ a 0-operator $A \in \Psi_0^0(X; {}^0\Omega^{1/2})$ such that ${}^{b,c}\sigma^{(0,0)}(\mathcal{N}_{\varrho_N}^{\nu}(A)(y, \widehat{\eta})) = 0$, $I_b^{(\mathfrak{a})}\left(\mathcal{N}_{\varrho_N}^{\nu}(A)(y, \widehat{\eta})\right) = 0$, and $I_c\left(\mathcal{N}_{\varrho_N}^{\nu}(A)(y, \widehat{\eta})\right)(\xi) = h_1(y, \widehat{\eta})h_2(\xi)$.

7.4. THE C^*-ALGEBRA GENERATED BY THE REDUCED NORMAL OPERATOR

Choose $\omega \in \mathcal{C}_c^\infty(\overline{\mathbb{R}}_+)$ with $\omega = 1$ near $x = 0$, and consider the function $a_0 : \overline{\mathbb{R}}_+ \times \mathbb{R}_y^{n-1} \times (\mathbb{R}_\xi \times \mathbb{R}_\eta^{n-1}) \setminus \{0\} \longrightarrow \mathbb{C}$ given by

$$a_0 : (x, y, \xi, \eta) \longmapsto \begin{cases} \omega(x) h_1(y, \eta/|\eta|) h_2(\xi/|\eta|) &, \eta \neq 0 \\ 0 &, \eta = 0 \end{cases}.$$

Then we have $a_0 \in S^{[0]}(\overline{\mathbb{R}}_+ \times \mathbb{R}_y^{n-1}; \mathbb{R}_\xi \times \mathbb{R}_\eta^{n-1})$, hence, for any excision function $\chi \in \mathcal{C}^\infty(\mathbb{R}_\xi \times \mathbb{R}_\eta^{n-1})$ with $\chi = 0$ near $(\xi, \eta) = 0$ and $\chi = 1$ for $|(\xi, \eta)|$ large, we get

$$[a : (x, y, \xi, \eta) \longmapsto \chi(\xi, \eta) a_0(x, y, \xi, \eta)] \in S_{cl}^0(\overline{\mathbb{R}}_+ \times \mathbb{R}_y^{n-1}; \mathbb{R}_\xi \times \mathbb{R}_\eta^{n-1})$$

with $\sigma^{(0)}(a) = a_0$. Let $A \in \Psi_0^0(X; {}^0\Omega^{1/2})$ be an operator whose lifted Schwartz kernel is near the boundary ∂X locally given by

$$\widehat{\kappa}_A(\tau, U, r, y') = \omega_0(\tau)^{Os{-}}\!\!\int_{\mathbb{R}_\eta^{n-1}} \int_{\mathbb{R}_\xi} e^{iU\eta} e^{i\tau\xi} a(r, y', \xi, \eta) d\xi d\eta,$$

where $\omega_0 \in \mathcal{C}_c^\infty((-1, 1))$ satisfies $\omega_0 = 1$ near $\tau = 0$. By Proposition 4.4.1 and Proposition 4.4.4, we have for $(y, \widehat{\eta}) \in \mathbb{R}_y^{n-1} \times S_{\widehat{\eta}}^{n-2}$

$$
\begin{aligned}
{}^{b,c}\sigma^{(0,0)}(\mathcal{N}_{\varrho_N}^\nu(A)(y, \widehat{\eta}))(z, \xi) &= a_0(0, y, \xi, 0) = 0 \text{ for } (z, \xi) \in {}^{b,c}T^*M \setminus \{0\} \\
I_c\left(\mathcal{N}_{\varrho_N}^\nu(A)(y, \widehat{\eta})\right)(\xi) &= a_0(0, y, \xi, \widehat{\eta}) = h_1(y, \widehat{\eta}) h_2(\xi) \text{ for } \xi \in \mathbb{R}_\xi \text{ and}
\end{aligned}
$$

$$I_b^{(\mathfrak{a})}\left(\mathcal{N}_{\varrho_N}^\nu(A)(y, \widehat{\eta})\right)(\xi - i\mathfrak{a}) = \sqrt{2}^{Os{-}}\!\!\int_{-1}^1 \int_{\mathbb{R}_\xi} \left(\frac{1+\tau}{1-\tau}\right)^{i\xi - \mathfrak{a}} e^{i\tau\xi} a(0, y, \xi, 0) d\xi \frac{d\tau}{(1-\tau^2)^{1/2}} = 0$$

for $\xi \in \mathbb{R}_\xi$.

(c) The inclusion (7.9) is a direct consequence of Proposition 4.4.4. Then the proof continues similarly to that of (b) but using the density of $M_\mathcal{O}^{-\infty}|_{\Gamma_\mathfrak{a}}$ in $\mathcal{C}_0(\Gamma_\mathfrak{a})$ [52, Corollary 3.3] instead of that of $\mathcal{C}_c^\infty(\mathbb{R}_\xi)$ in $\mathcal{C}_0(\mathbb{R}_\xi)$ as in (b)

(d) It remains to show $J_2 \subseteq \mathcal{B}_{b,c}^{(\mathfrak{a})}(M, {}^{b,c}\Omega^{1/2}; S^*\partial X)$. Let $\mathcal{N} \in J_2$ be arbitrary. By (b) and (c) there exists $A \in \mathcal{B}_0^{(\mathfrak{a})}(X, {}^0\Omega^{1/2})$ with ${}^{b,c}\sigma^{(0,0)}(\mathcal{N}_{\varrho_N}^{\nu,\mathfrak{a}}(A)) = 0$, $I_b^{(\mathfrak{a})}\left(\mathcal{N}_{\varrho_N}^{\nu,\mathfrak{a}}(A)\right) = I_b^{(\mathfrak{a})}(\mathcal{N})$, and $I_c\left(\mathcal{N}_{\varrho_N}^{\nu,\mathfrak{a}}(A)\right) = I_c(\mathcal{N})$, hence we have

$$\mathcal{N} - \mathcal{N}_{\varrho_N}^{\nu,\mathfrak{a}}(A) \in \mathcal{C}(S^*\partial X, \mathcal{K}(\varrho_0^{\mathfrak{a}} L^2(M, {}^{b,c}\Omega^{1/2}))) = J_1 \subseteq \mathcal{B}_{b,c}^{(\mathfrak{a})}(M, {}^{b,c}\Omega^{1/2}; S^*\partial X)$$

by (3.14), and therefore $\mathcal{N} \in \mathcal{B}_{b,c}^{(\mathfrak{a})}(M, {}^{b,c}\Omega^{1/2}; S^*\partial X)$ by (a). $\square$

PROPOSITION 7.4.3. *The C^*-algebra $\mathcal{B}_{b,c}^{(\mathfrak{a})}(M, {}^{b,c}\Omega^{1/2}; S^*\partial X)$ is a solvable C^*-algebra of length 1. A solving composition series is given by*

(7.10) $$0 \subseteq J_1 \subseteq J_2 \subseteq \mathcal{B}_{b,c}^{(\mathfrak{a})}(M, {}^{b,c}\Omega^{1/2}; S^*\partial X).$$

The subquotients are determined by

(7.11) $\mathcal{B}_{b,c}^{(\mathfrak{a})}(M, {}^{b,c}\Omega^{1/2}; S^*\partial X)/J_2 \cong \mathcal{C}(\partial X) \oplus \mathcal{C}(\partial X)$

(7.12) $\quad\quad\quad\quad\quad\quad\quad\quad J_2/J_1 \cong \mathcal{C}_0(S^*\partial X \times \mathbb{R}) \oplus \mathcal{C}_0(\partial X \times \Gamma_\mathfrak{a})$ *and*

$$J_1 = \mathcal{C}(S^*\partial X, \mathcal{K}(\varrho_0^{\mathfrak{a}} L^2(M, {}^{b,c}\Omega^{1/2}))).$$

Proof: Because of Lemma 7.4.2 it remains to establish the isomorphisms (7.11) and (7.12). Consider

$$F_1 : \mathcal{B}_{b,c}^{(\mathfrak{a})}(M, {}^{b,c}\Omega^{1/2}; S^*\partial X) \longrightarrow \mathcal{C}(\partial X) \oplus \mathcal{C}(\partial X):$$
$$\mathcal{N} \longmapsto {}^{b,c}\sigma^{(0,0)}(\mathcal{N}) \text{ , and}$$
$$F_2 : J_2 \longrightarrow \mathcal{C}_0(S^*\partial X \times \mathbb{R}) \oplus \mathcal{C}_0(\partial X \times \Gamma_{\mathfrak{a}}):$$
$$\mathcal{N} \longmapsto \left(I_c(\mathcal{N}), I_b^{(\mathfrak{a})}(\mathcal{N})\right).$$

By Lemma 7.4.2, F_2 is onto with kernel J_1 thus, we get (7.12). As for (7.11), given any $h \in \mathcal{C}^\infty(\partial X) \oplus \mathcal{C}^\infty(\partial X)$, we are going to show that there exists a 0-operator $A \in \Psi_0^0(X; {}^0\Omega^{1/2})$ with $F_1(\mathcal{N}_{\varrho N}^\nu(A)) = h$. By a partition of unity, we can assume $h \in \mathcal{C}_c^\infty(\mathbb{R}_y^{n-1} \times \{\pm 1\})$. Let $a \in S_{cl}^0(\overline{\mathbb{R}}_+ \times \mathbb{R}_y^{n-1}; \mathbb{R}_\xi \times \mathbb{R}_\eta^{n-1})$ be with $\sigma^{(0)}(a)(0, y, \pm 1, 0) = h(y, \pm 1)$. Then the operator $A \in \Psi_0^0(X; {}^0\Omega^{1/2})$ whose lifted Schwartz kernel is given near the boundary by

$$\hat{\kappa}_A(\tau, U, r, y') = \omega(\tau, U) \mathrm{Os}{-}\!\!\int_{\mathbb{R}_\eta^{n-1}} \int_{\mathbb{R}_\xi} e^{iU\eta} e^{i\tau\xi} a(r, y', \xi, \eta) d\xi d\eta$$

for some $\omega \in \mathcal{C}_c^\infty((-1,1) \times \mathbb{R}_U^{n-1})$ with $\omega = 1$ near $(\tau, U) = 0$ has the desired properties by Proposition 4.4.4. Consequently, F_1 is onto with kernel J_1, therefore, we obtain (7.11).

Since $\mathcal{B}_{b,c}^{(\mathfrak{a})}(M, {}^{b,c}\Omega^{1/2}; S^*\partial X)/J_1$ is a commutative C^*-algebra, the C^*-algebra $\mathcal{B}_{b,c}^{(\mathfrak{a})}(M, {}^{b,c}\Omega^{1/2}; S^*\partial X)$ is solvable of length at most one; on the other hand, by (7.11) and (7.12), we see that $\mathcal{B}_{b,c}^{(\mathfrak{a})}(M, {}^{b,c}\Omega^{1/2}; S^*\partial X)$ has irreducible representations of different dimensions (1 and ∞), hence, it is solvable of length at least 1. This completes the proof. □

7.5. The C^*-algebra $\mathcal{B}_0^{(\mathfrak{a})}(X, {}^0\Omega^{1/2})$

Let us start with the characterization of the C^*-algebra $Q_0^{(\mathfrak{a})}$.

PROPOSITION 7.5.1. *The C^*-algebra $Q_0^{(\mathfrak{a})}$ of joint 0-symbols consists of all pairs*

$$(f, \mathcal{N}) \in \mathcal{C}({}^0S^*X) \oplus \mathcal{B}_{b,c}^{(\mathfrak{a})}(M, {}^{b,c}\Omega^{1/2}; S^*\partial X) \subseteq \mathcal{C}({}^0S^*X) \oplus \mathcal{C}(S^*\partial X, \mathcal{B}_{b,c}^{(\mathfrak{a})}(M, {}^{b,c}\Omega^{1/2}))$$

satisfying the compatibility conditions (7.5) and (7.6).

Proof: Let us denote by Q' the space described in the Proposition. Then Q' is a C^*-algebra that contains $\tau_0(\Psi_0^0(X; {}^0\Omega^{1/2}))$ by Proposition 4.4.4, thus, we have $Q_0^{(\mathfrak{a})} \subseteq Q'$. On the other hand, let $(f, \mathcal{N}) \in Q'$ be arbitrary. Since the map ${}^0\sigma^{(0)} : \mathcal{B}_0^{(\mathfrak{a})}(X, {}^0\Omega^{1/2}) \to \mathcal{C}({}^0S^*X)$ is onto, there exists $A_0 \in \mathcal{B}_0^{(\mathfrak{a})}(X, {}^0\Omega^{1/2})$ with ${}^0\sigma^{(0)}(A_0) = f$. By Proposition 7.3.2, (7.5) and (7.6) ${}^{b,c}\sigma^{(0,0)}(\mathcal{N}_{\varrho N}^{\nu,\mathfrak{a}}(A_0) - \mathcal{N}) = 0$ and $I_c\left(\mathcal{N}_{\varrho N}^{\nu,\mathfrak{a}}(A_0) - \mathcal{N}\right) = 0$, thus, we get $I_b^{(\mathfrak{a})}\left(\mathcal{N}_{\varrho N}^{\nu,\mathfrak{a}}(A_0) - \mathcal{N}\right) =: h \in \mathcal{C}_0(\partial X \times \Gamma_{\mathfrak{a}})$. By Lemma 7.4.2, there exists $A_1 \in N({}^0\sigma^{(0)})$ with $I_b^{(\mathfrak{a})}\left(\mathcal{N}_{\varrho N}^{\nu,\mathfrak{a}}(A_1)\right) = h$, hence $\mathcal{N}_{\varrho N}^{\nu,\mathfrak{a}}(A_0 - A_1) - \mathcal{N} \in \mathcal{C}(S^*\partial X, \mathcal{K}(\varrho_0^\mathfrak{a} L^2(M, {}^{b,c}\Omega^{1/2})))$, and we obtain $A_2 \in N({}^0\sigma^{(0)})$ with $\mathcal{N}_{\varrho N}^{\nu,\mathfrak{a}}(A_2) = \mathcal{N}_{\varrho N}^{\nu,\mathfrak{a}}(A_0 - A_1) - \mathcal{N}$ by Lemma 7.4.2. Summarizing, the operator $A := A_0 - A_1 - A_2 \in \mathcal{B}_0^{(\mathfrak{a})}(X, {}^0\Omega^{1/2})$ satisfies $\tau_0(A) = ({}^0\sigma^{(0)}(A), \mathcal{N}_{\varrho N}^{\nu,\mathfrak{a}}(A)) = (f, \mathcal{N})$. This completes the proof. □

Consider now the following closed, two-sided ideals in $Q_0^{(a)}$.

$$\mathcal{J}_1 := \{(0,\mathcal{N}) \in Q_0^{(a)} : I_b(\mathcal{N}(\widehat{\eta})) = 0 \text{ for all } \widehat{\eta} \in S^*\partial X\} \text{ and}$$
$$\mathcal{J}_2 := \{(f,\mathcal{N}) \in Q_0^{(a)} : f = 0\}.$$

PROPOSITION 7.5.2. *The C^*-algebra $Q_0^{(a)}$ is solvable of length 1. A solving composition series is given by*

$$0 \subseteq \mathcal{J}_1 \subseteq \mathcal{J}_2 \subseteq Q_0^{(a)}.$$

The subquotients are determined by the isomorphisms

(7.13) $\qquad\qquad Q_0^{(a)}/\mathcal{J}_2 \cong \mathcal{C}(^0S^*X)$
(7.14) $\qquad\qquad \mathcal{J}_2/\mathcal{J}_1 \cong \mathcal{C}_0(\partial X \times \Gamma_\mathfrak{a})$ *and*
(7.15) $\qquad\qquad \mathcal{J}_1 \cong \mathcal{C}(S^*\partial X, \mathcal{K}(\varrho_0^\mathfrak{a} L^2(M, {}^{b,c}\Omega^{1/2}))).$

Proof: As in the proof of Proposition 7.4.3, it remains to establish the isomorphisms (7.13), (7.14), and (7.15). Consider

$$F_1 : Q^{(a)} \longrightarrow \mathcal{C}(^0S^*X) : (f,\mathcal{N}) \longmapsto f,$$
$$F_2 : \mathcal{J}_2 \longrightarrow \mathcal{C}_0(\partial X \times \Gamma_\mathfrak{a}) : (0,\mathcal{N}) \longmapsto \left[(\widehat{\eta},\xi) \mapsto I_b^{(a)}(\mathcal{N}(\widehat{\eta}))(\xi)\right], \text{ and}$$
$$F_3 : \mathcal{J}_1 \longrightarrow \mathcal{C}(S^*\partial X, \mathcal{K}(\varrho_0^\mathfrak{a} L^2(M, {}^{b,c}\Omega^{1/2}))) : (0,\mathcal{N}) \longmapsto \mathcal{N}.$$

Then F_1 is onto with kernel $\mathcal{J}_2$ by Remark 7.3.1, F_2 is onto with kernel $\mathcal{J}_1$ by Lemma 7.4.2, and F_3 is one-to-one and onto again by Lemma 7.4.2. The computation of the length of $Q_0^{(a)}$ proceeds exactly as that of $\mathcal{B}_{b,c}^{(a)}(M, {}^{b,c}\Omega^{1/2}; S^*\partial X)$ in Proposition 7.4.3. $\square$

COROLLARY 7.5.3. *The C^*-algebra $\mathcal{B}_0^{(a)}(X, {}^0\Omega^{1/2})$ is solvable of length 2. A solving composition series is given by*
(7.16)
$$0 \subseteq \mathcal{K}(\varrho_N^\mathfrak{a} L^2(X, {}^0\Omega^{1/2})) \subseteq \mathcal{I}_2 := \tau_0^{-1}(\mathcal{J}_1) \subseteq \mathcal{I}_3 := \tau_0^{-1}(\mathcal{J}_2) \subseteq \mathcal{B}_0^{(a)}(X, {}^0\Omega^{1/2}),$$

where $\mathcal{J}_k$, $k = 1,2$ is the ideal of Proposition 7.5.2. The subquotients are determined by

$$\mathcal{B}_0^{(a)}(X, {}^0\Omega^{1/2})/\mathcal{I}_3 \cong \mathcal{C}(^0S^*X)$$
$$\mathcal{I}_3/\mathcal{I}_2 \cong \mathcal{C}_0(\partial X \times \Gamma_\mathfrak{a}) \text{ and}$$
$$\mathcal{I}_2/\mathcal{K}(\varrho_N^\mathfrak{a} L^2(X, {}^0\Omega^{1/2})) \cong \mathcal{C}(S^*\partial X, \mathcal{K}(\varrho_0^\mathfrak{a} L^2(M, {}^{b,c}\Omega^{1/2}))).$$

Proof: This is an immediate consequence of Proposition 7.5.2. For the computation of the length use for instance [42, Lemma 3.1]. $\square$

7.6. The spectrum of the C^*-algebra $\mathcal{B}_0^{(a)}(X, {}^0\Omega^{1/2})$

Since the C^*-algebra $\mathcal{B}_0^{(a)}(X, {}^0\Omega^{1/2})$ is solvable, the spectrum $\widehat{\mathcal{B}_0^{(a)}(X, {}^0\Omega^{1/2})}$ is determined as a set by the composition series (7.16) in Corollary 7.5.3. We are going to determine the Jacobson topology on the spectrum $\widehat{\mathcal{B}_0^{(a)}(X, {}^0\Omega^{1/2})}$. In fact, it suffices to compute the Jacobson topology on the spectrum $\widehat{Q_0^{(a)}}$ of the C^*-algebra $Q_0^{(a)}$ of joint 0-symbols.

PROPOSITION 7.6.1.   (a) Let $\pi : \mathcal{B}_0^{(\mathfrak{a})}(X,{}^0\Omega^{1/2}) \to \mathcal{L}(H_\pi)$ be an irreducible representation. Then either π is equivalent to the identity representation $\mathrm{id} : \mathcal{B}_0^{(\mathfrak{a})}(X,{}^0\Omega^{1/2}) \to \mathcal{L}(\varrho_N^\mathfrak{a} L^2(X,{}^0\Omega^{1/2}))$ or there exists an irreducible representation $\varrho : Q_0^{(\mathfrak{a})} \to \mathcal{L}(H_\varrho)$ such that π and $\varrho \circ \tau_0$ are equivalent. Moreover, $\varrho_1 \circ \tau_0$ and $\varrho_2 \circ \tau_0$ are equivalent if and only if ϱ_1 and ϱ_2 are equivalent as representations of $Q_0^{(\mathfrak{a})}$.

(b) The Jacobson topology $\mathcal{T}_{\mathcal{J}}(\mathcal{B}_0^{(\mathfrak{a})}(X,{}^0\Omega^{1/2}))$ on the spectrum $\widehat{\mathcal{B}_0^{(\mathfrak{a})}(X,{}^0\Omega^{1/2})}$ is given by

$$\mathcal{T}_{\mathcal{J}}(\mathcal{B}_0^{(\mathfrak{a})}(X,{}^0\Omega^{1/2})) = \{\emptyset\} \cup \left\{ \{[\mathrm{id}]\} \cup \{[\varrho \circ \tau_0] : [\varrho] \in V\} : V \subseteq \widehat{Q_0^{(\mathfrak{a})}} \text{ open} \right\}$$

provided the spectrum $\widehat{Q_0^{(\mathfrak{a})}}$ is equipped with the Jacobson topology.

Proof: This is an immediate consequence of [15, Proposition 2.11.2, Proposition 3.2.1] and the exactness of (7.4).   □

Let us therefore consider the spectrum of the algebra $Q_0^{(\mathfrak{a})}$. By (7.2), the spectrum $\widehat{Q_0^{(\mathfrak{a})}}$ of $Q_0^{(\mathfrak{a})}$ is determined as a set by the composition series in Proposition 7.5.2. To be more precise, let

$$\pi_\zeta : Q_0^{(\mathfrak{a})} \longrightarrow \mathbb{C} : (f,\mathcal{N}) \longmapsto f(\zeta), \text{ for } \zeta \in {}^0S^*X$$

$$\pi_{q,\lambda} : Q_0^{(\mathfrak{a})} \longrightarrow \mathbb{C} : (f,\mathcal{N}) \longmapsto I_b^{(\mathfrak{a})}(\mathcal{N}(\widehat{\eta}))(\lambda), \text{ for } \lambda \in \Gamma_\mathfrak{a}, \pi(\widehat{\eta}) = q, \text{ and}$$

$$\pi_{\widehat{\eta}} : Q_0^{(\mathfrak{a})} \longrightarrow \mathcal{L}(\varrho_0^\mathfrak{a} L^2(M,{}^{b,c}\Omega^{1/2})) : (f,\mathcal{N}) \longmapsto \mathcal{N}(\widehat{\eta}), \text{ for } \widehat{\eta} \in S^*\partial X.$$

PROPOSITION 7.6.2. The representations π_ζ, $\zeta \in {}^0S^*X$, $\pi_{q,\lambda}$, $(q,\lambda) \in \partial X \times \Gamma_\mathfrak{a}$, and $\pi_{\widehat{\eta}}$, $\widehat{\eta} \in S^*\partial X$, of $Q_0^{(\mathfrak{a})}$ are irreducible and pairwise inequivalent. Moreover, any irreducible representation of $Q_0^{(\mathfrak{a})}$ is equivalent to one of them. In particular, we obtain a bijective map

(7.17) $\qquad \Phi : \widehat{Q_0^{(\mathfrak{a})}} \longrightarrow \mathfrak{M} := {}^0S^*X \uplus (\partial X \times \Gamma_\mathfrak{a}) \uplus S^*\partial X : [\pi_\mathfrak{m}] \longmapsto \mathfrak{m}.$

7.6.3. A topology on $\mathfrak{M}$. We are going to describe a topology $\mathcal{T}$ on $\mathfrak{M}$ by characterizing local bases $\mathfrak{U}(\mathfrak{m})$ at the points $\mathfrak{m} \in \mathfrak{M}$.

In case $\zeta \in {}^0S^*X$ with $\pi(\zeta) \notin \partial X$, let $\mathfrak{U}(\zeta)$ be the family of all open sets $U \subseteq {}^0S^*X$ with $\zeta \in U$ and $\pi(U) \cap \partial X = \emptyset$. For $\zeta \in {}^0S^*X$ with $\pi(\zeta) \in \partial X$ let $\omega \in {}^0T^*X \setminus \{0\}$ be a lift of ζ, i.e. $\widehat{\omega} = \zeta$. Recall furthermore that by (4.18) the choice of a boundary defining function $\varrho_N : X \to \overline{\mathbb{R}}_+$ yields an identification $j_{\varrho_N} : T^*\partial X \oplus \partial X \times \mathbb{R} \xrightarrow{\cong} {}^0T^*X|_{\partial X}$. Let $\eta \in T_q^*\partial X$ and $\lambda = \xi - i\mathfrak{a} \in \Gamma_\mathfrak{a}$ be with $j_{\varrho_N}(\eta,(q,\xi)) = \omega$. If $\eta \neq 0$, let $\widehat{\eta} \in S^*\partial X$ be the projection of η in $S^*\partial X$, and let $\mathfrak{U}(\zeta)$ be the set of all $U \uplus W$ where $U \subseteq {}^0S^*X$ resp. $W \subseteq S^*\partial X$ are open with $\zeta \in U$ resp. $\widehat{\eta} \in W$, and $\eta' \neq 0$ for all $j_{\varrho_N}(\eta',(q',\xi')) = \omega'$ with $\zeta' = \widehat{\omega'} \in U$. In case $\eta = 0$, let $\mathfrak{U}(\zeta)$ be the set of all sets of the form

$$U \uplus (V \times \mathcal{W}_c) \uplus \pi^{-1}(V)$$

where $U \subseteq {}^0S^*X$ resp. $V \subseteq \partial X$ are open with $\zeta \in U$ resp. $q \in V$, and $\mathcal{W}_c \subseteq \Gamma_\mathfrak{a}$ is an open conical set such that $t\lambda := t\xi - i\mathfrak{a} \in \mathcal{W}_c$ for some $t > 0$.

For $\mathfrak{m} = (q,\lambda) \in \partial X \times \Gamma_\mathfrak{a}$, let $\mathfrak{U}(\mathfrak{m})$ be the set of all of sets of the form $(V \times \mathcal{W}) \uplus \pi^{-1}(V)$ where $V \subseteq \partial X$ resp. $\mathcal{W} \subseteq \Gamma_\mathfrak{a}$ are open with $q \in V$ resp. $\lambda \in \mathcal{W}$.

7.6. THE SPECTRUM OF THE C^*-ALGEBRA $\mathcal{B}_0^{(\mathfrak{a})}(X, {}^0\Omega^{1/2})$ 65

Finally, for $\widehat{\eta} \in S^*\partial X$, let $\mathfrak{U}(\widehat{\eta})$ be the set of all open sets $W \subseteq S^*\partial X$ with $\widehat{\eta} \in W$.

Then the systems $\mathfrak{U}(\mathfrak{m})$, $\mathfrak{m} \in \mathfrak{M}$, satisfy the conditions (1), (2) and (3) of [**77**, p. 41], hence, there exists a unique topology $\mathcal{T}$ on $\mathfrak{M}$ such that $\mathfrak{U}(\mathfrak{m})$ is a local base at $\mathfrak{m}$ for all $\mathfrak{m} \in \mathfrak{M}$.

Before proving that the topology $\mathcal{T}$ on $\mathfrak{M}$ coincides with the Jacobson topology, we need the following Lemmata. The notations are as in 7.6.3.

LEMMA 7.6.4. *Let $\zeta_0 \in {}^0S^*X|_{\partial X}$ and $\omega_0 = j_{\varrho_N}(\eta_0, (q_0, \xi_0)) \in {}^0T^*X \setminus \{0\}$ a lift of ζ_0.*

(a) *If $\eta_0 \neq 0$, then for any $\mathcal{U}_0 = U \uplus W \in \mathfrak{U}(\zeta_0)$ there exists $A \in \mathcal{B}_0^{(\mathfrak{a})}(X, {}^0\Omega^{1/2})$ such that $|{}^0\sigma^{(0)}(A)(\zeta_0)| > 1$, $\mathrm{supp}\,{}^0\sigma^{(0)}(A) \subseteq U$, $I_b^{(\mathfrak{a})}\left(\mathcal{N}_{\varrho_N}^{\nu,\mathfrak{a}}(A)\right) = 0$, and $\mathrm{supp}\,\mathcal{N}_{\varrho_N}^{\nu,\mathfrak{a}}(A) \subseteq W$.*

(b) *If $\eta_0 = 0$, then for any $\mathcal{U}_0 = U \uplus (V \times \mathcal{W}_c) \uplus \pi^{-1}(V)$ there exists a bounded operator $A \in \mathcal{B}_0^{(\mathfrak{a})}(X, {}^0\Omega^{1/2})$ such that we have $|{}^0\sigma^{(0)}(A)(\zeta_0)| > 1$, $\mathrm{supp}\,{}^0\sigma^{(0)}(A) \subseteq U$, $I_b^{(\mathfrak{a})}\left(\mathcal{N}_{\varrho_N}^{\nu,\mathfrak{a}}(A)\right)(\widehat{\eta}, \lambda) = 0$ for all $(\widehat{\eta}, \lambda) \notin V \times \mathcal{W}_c$, and $\mathrm{supp}\,\mathcal{N}_{\varrho_N}^{\nu,\mathfrak{a}}(A) \subseteq \pi^{-1}(V)$.*

Proof: Let us choose local coordinates near $q_0 \in \partial X$ such that q_0 corresponds to $y_0 \in \mathbb{R}_y^{n-1}$, and η_0 to $(y_0, \eta_0) \in \mathbb{R}_y^{n-1} \times \mathbb{R}_\eta^{n-1}$, i.e. ζ_0 corresponds to $(0, y_0, \xi_0, \eta_0)$. Without loss of generality, we can further assume $|\xi_0|^2 + |\eta_0|^2 = 1$ and that $s(S^*\partial X) \subseteq T^*\partial X \setminus \{0\}$ corresponds to $\mathbb{R}_y^{n-1} \times S_{\widehat{\eta}}^{n-2}$.

(a) Choose $\varepsilon_0 > 0$ so small that

$$W_1 := \left\{ y \in \mathbb{R}_y^{n-1} : |y - y_0| < \varepsilon_0 \right\} \times \left\{ \widehat{\eta} \in S_{\widehat{\eta}}^{n-2} : |\widehat{\eta} - \eta_0/|\eta_0|| < \varepsilon_0 \right\}$$

$$\subseteq \left\{ y \in \mathbb{R}_y^{n-1} : |y - y_0| < 2\varepsilon_0 \right\} \times \left\{ \widehat{\eta} \in S_{\widehat{\eta}}^{n-2} : |\widehat{\eta} - \eta_0/|\eta_0|| < 2\varepsilon_0 \right\} \subseteq W,$$

and let $V_1 := \{t\widehat{\eta} \in \mathbb{R}_\eta^{n-1} : 0 < t \leq 1 \text{ and } |\widehat{\eta} - \eta_0/|\eta_0|| < \varepsilon_0\}$. Then $U_1 := \{(\xi, \eta) \in S^{n-1} : \eta \in V_1\}$ is open in S^{n-1} with $(\xi_0, \eta_0) \in U_1$, thus, there exists $\varepsilon_0 > \varepsilon_1 > 0$ with

$$U_2 := \{(\xi, \eta) \in S^{n-1} : |\xi - \xi_0| < \varepsilon_1 \text{ and } |\eta - \eta_0| < \varepsilon_1\} \subseteq U_1,$$

$\{(x, y) \in \overline{\mathbb{R}}_+ \times \mathbb{R}_y^{n-1} : x < \varepsilon_1 \text{ and } |y - y_0| < \varepsilon_1\} \times U_2 \subseteq U \subseteq {}^0S^*X$, and $(\pm 1, 0) \notin U_2$, where we have used the defining function ϱ_N, the normal fibration ν, and the local coordinates on ∂X to trivialize ${}^0T^*X$ near the boundary ∂X.

Choose $\varphi \in \mathcal{C}_c^\infty(\mathbb{R}_y^{n-1})$ with $\varphi(y_0) = 2$ and $\mathrm{supp}\,\varphi \subseteq \{|y - y_0| < \varepsilon_1\}$, $\psi \in \mathcal{C}_c^\infty(U_2)$ with $\psi(\xi_0, \eta_0) = 2$, $\omega \in \mathcal{C}_c^\infty(\overline{\mathbb{R}}_+)$ with $\omega = 1$ near $r = 0$, $\mathrm{supp}\,\omega \subseteq \{x < \varepsilon_1\}$, and $\chi \in \mathcal{C}^\infty(\mathbb{R}_\xi \times \mathbb{R}_\eta^{n-1})$ such that $\chi = 0$ for $|(\xi, \eta)| < 1$ and $\chi = 1$ for $|(\xi, \eta)| > 2$. Then we have

$$\left[a : (x, y, \xi, \eta) \longmapsto \omega(x)\varphi(y)\chi(\xi, \eta)\psi\left(\frac{(\xi, \eta)}{|(\xi, \eta)|}\right) \right] \in S_{cl}^0(\overline{\mathbb{R}}_+ \times \mathbb{R}_y^{n-1}; \mathbb{R}_\xi \times \mathbb{R}_\eta^{n-1}).$$

Let now $\omega_0 \in \mathcal{C}_c^\infty((-1, 1))$ be with $\omega_0 = 1$ near $\tau = 0$ and consider the operator $A_1 \in \Psi_0^0(X; {}^0\Omega^{1/2})$ whose lifted Schwartz kernel is given by

$$\widehat{\kappa}_{A_1}(\tau, U, r, y') = \omega_0(\tau) {}^{Os\text{-}}\!\!\int_{\mathbb{R}_\eta^{n-1}} \int_{\mathbb{R}_\xi} e^{i\tau\xi} e^{iU\eta} a(r, y', \xi, \eta) d\xi d\eta.$$

Then $\operatorname{supp}{}^0\sigma^{(0)}(A_1) \subseteq U$ and $\operatorname{supp}\mathcal{N}_{\varrho_N}^\nu(A_1) \subseteq W_1$ by a straightforward computation using (4.23). Let $h_1 := I_b^{(\mathfrak{a})}\left(\mathcal{N}_{\varrho_N}^\nu(A_1)\right) \in \mathcal{C}(\partial X, \mathcal{B}(;\Gamma_\mathfrak{a}))$. Then the compatibility conditions (7.5) and (7.8) yield for all $\widehat{\eta} \in \pi^{-1}(q_0)$

$$\sigma_\mathcal{B}^{(0)}(h_1)(q_0, \pm 1) = {}^{b,c}\sigma^{(0,0)}(\mathcal{N}_{\varrho_N}^\nu(A_1)(\widehat{\eta}))(0, \pm 1) = {}^0\sigma^{(0)}(A)_0(j_{\varrho_N}(0,)(q, \pm 1)) = 0$$

because of $(\pm 1, 0) \notin U_2$, thus, we have $h_1 \in \mathcal{C}_0(\partial X \times \Gamma_\mathfrak{a})$ and there exists $A_2 \in N({}^0\sigma^{(0)})$ with $I_b^{(\mathfrak{a})}\left(\mathcal{N}_{\varrho_N}^{\nu,\mathfrak{a}}(A_2)\right) = h_1$ by Lemma 7.4.2. Let $A_3 := A_1 - A_2$, choose $\vartheta \in \mathcal{C}(S^*\partial X)$ with $\vartheta|_{W_1} = 0$ and $\vartheta(y, \widehat{\eta}) = 1$ for all $(y, \widehat{\eta})$ with $|y - y_0| > 2\varepsilon_0$ and $|\widehat{\eta} - \eta_0/|\eta_0|| > 2\varepsilon_0$, and consider the function $\mathcal{N} := \vartheta\mathcal{N}_{\varrho_N}^{\nu,\mathfrak{a}}(A_3) \in \mathcal{C}(S^*\partial X, \mathcal{B}_{b,c}^{(\mathfrak{a})}(M, {}^{b,c}\Omega^{1/2}))$. Because of

$$\begin{aligned}
{}^{b,c}\sigma^{(0,0)}(\mathcal{N}(\widehat{\eta}))(z, \xi) &= \vartheta(\widehat{\eta})^{b,c}\sigma^{(0,0)}(\mathcal{N}_{\varrho_N}^\nu(A_1)(\widehat{\eta}))(z, \xi) = 0 \\
I_b^{(\mathfrak{a})}\left(\mathcal{N}(\widehat{\eta})\right) &= \vartheta(\widehat{\eta})\left(\mathcal{N}_{\varrho_N}^\nu(A_1)(\widehat{\eta}) - \mathcal{N}_{\varrho_N}^{\nu,\mathfrak{a}}(A_2)(\widehat{\eta})\right) = 0 \text{ and} \\
I_c\left(\mathcal{N}(\widehat{\eta})\right) &= \vartheta(\widehat{\eta})I_c\left(\mathcal{N}_{\varrho_N}^\nu(A_1)(\widehat{\eta})\right) = 0
\end{aligned}$$

by $\operatorname{supp}\mathcal{N}_{\varrho_N}^\nu(A_1) \subseteq W_1$ and the choice of ϑ, we obtain

$$\mathcal{N} \in \mathcal{C}(S^*\partial X, \mathcal{K}(\varrho_0^\mathfrak{a} L^2(M, {}^{b,c}\Omega^{1/2}))) = J_1.$$

By Lemma 7.4.2, there exists $A_4 \in N({}^0\sigma^{(0)})$ with $\mathcal{N}_{\varrho_N}^{\nu,\mathfrak{a}}(A_4) = \mathcal{N}$. Then $A := A_3 - A_4$ has the desired properties.

(b) Because of $\eta_0 = 0$ we have $\xi_0 \neq 0$. Without loss of generality we can assume $\xi_0 > 0$, thus, there exists $R > 0$ with $[R, \infty) \times \{-i\mathfrak{a}\} \subseteq \mathcal{W}_c$. Let $f \in \mathcal{C}_c^\infty(U)$ be with $f(\zeta_0) = 2$ and $f(-\zeta_0) = 0$ where a lift of $-\zeta_0$ is given by $j_{\varrho_N}(0, (q_0, -\xi_0)) \in T^*\partial X \setminus \{0\}$. Let $A_1 \in \Psi_0^0(X; {}^0\Omega^{1/2})$ be with ${}^0\sigma^{(0)}(A_1) = f$. Choose $\chi \in \mathcal{C}^\infty(\Gamma_\mathfrak{a})$ with $\chi(\xi - i\mathfrak{a}) = 0$ for $\xi \leq 2R$ and $\chi(\xi - i\mathfrak{a}) = 1$ for $\xi \geq 3R$. Then the function

$$h : (q, \lambda) \mapsto \sigma_\mathcal{B}^{(0)}\left(I_b\left(\mathcal{N}_{\varrho_N}^\nu(A_1)(\widehat{\eta})\right)\right)(+1)\chi(\lambda)$$

with $\pi(\widehat{\eta}) = q$ is well-defined, belongs to $\mathcal{C}(\partial X, \mathcal{B}(;\Gamma_\mathfrak{a}))$ and satisfies $\sigma_\mathcal{B}^{(0)}(h(q))(\pm 1) = \sigma_\mathcal{B}^{(0)}(I_b\left(\mathcal{N}_{\varrho_N}^\nu(A_1)(\widehat{\eta})\right)(\pm 1)$, therefore,

$$I_b\left(\mathcal{N}_{\varrho_N}^\nu(A_1)\right) - h \in \mathcal{C}_0(\partial X \times \Gamma_\mathfrak{a}),$$

and there exists $A_2 \in N({}^0\sigma^{(0)})$ with $I_b\left(\mathcal{N}_{\varrho_N}^{\nu,\mathfrak{a}}(A_2)\right) = I_b\left(\mathcal{N}_{\varrho_N}^\nu(A_1)\right) - h$ by Lemma 7.4.2. Furthermore, choose $\varphi_b \in \mathcal{C}_c^\infty(V)$ and $\varphi \in \mathcal{C}^\infty(X)$ with $\varphi|_{\partial X} = \varphi_b$ and $\varphi = 1$ near $q_0 \in \partial X$. Then $A := M_\varphi(A_1 - A_2)$ has the desired properties by Example 4.4.5. $\square$

LEMMA 7.6.5. *Let* $\mathfrak{m} = (q_0, \lambda_0) \in \partial X \times \Gamma_\mathfrak{a}$ *be arbitrary. Then for any open neighborhood* $\mathcal{U}_0 = (V \times \mathcal{W}) \uplus \pi^{-1}(V) \in \mathfrak{U}(\mathfrak{m})$, *there exists* $A \in \mathcal{B}_0^{(\mathfrak{a})}(X, {}^0\Omega^{1/2})$ *such that* ${}^0\sigma^{(0)}(A) = 0$, $|I_b^{(\mathfrak{a})}\left(\mathcal{N}_{\varrho_N}^{\nu,\mathfrak{a}}(A)(\widehat{\eta})\right)(\lambda_0)| > 1$ *for all* $\widehat{\eta} \in \pi^{-1}(q_0) \subseteq S^*\partial X$, $I_b^{(\mathfrak{a})}\left(\mathcal{N}_{\varrho_N}^{\nu,\mathfrak{a}}(A)(\widehat{\eta})\right)(\lambda) = 0$ *for all* $(\widehat{\eta}, \lambda) \notin \pi^{-1}(V) \times \Gamma_\mathfrak{a}$, *and* $\operatorname{supp}\mathcal{N}_{\varrho_N}^{\nu,\mathfrak{a}}(A) \subseteq \pi^{-1}(V)$.

Proof: Let $h \in \mathcal{C}_c(V \times \mathcal{W})$ be arbitrary with $h(q_0, \lambda_0) = 2$, and using Lemma 7.4.2 choose $A_1 \in N({}^0\sigma^{(0)})$ with $I_b^{(\mathfrak{a})}\left(\mathcal{N}_{\varrho_N}^{\nu,\mathfrak{a}}(A_1)\right) = h$. As in the proof of Lemma 7.6.4, let $\varphi_b \in \mathcal{C}^\infty(\partial X)$ and $\varphi \in \mathcal{C}^\infty(X)$ be with $\varphi|_{\partial X} = \varphi_b$ and $\varphi = 1$ near $q_0 \in \partial X$. Then $A := M_\varphi A_1 \in \mathcal{B}_0^{(\mathfrak{a})}(X, {}^0\Omega^{1/2})$ has the desired properties. $\square$

We can now state and prove the main result of this subsection.

7.6. THE SPECTRUM OF THE C^*-ALGEBRA $\mathcal{B}_0^{(\mathfrak{a})}(X, {}^0\Omega^{1/2})$

THEOREM 7.6.6. *The bijective map* $\Phi : \widehat{Q_0^{(\mathfrak{a})}} \to \mathfrak{M}$ *of (7.17) is a homeomorphism provided* $\widehat{Q_0^{(\mathfrak{a})}}$ *is endowed with the Jacobson topology and* $\mathfrak{M}$ *with the topology* $\mathcal{T}$ *described in 7.6.3.*

Proof: Throughout the proof, let $\mathfrak{V}_q := \{[\pi] \in \widehat{Q_0^{(\mathfrak{a})}} : \|\pi(q)\|_{\mathcal{L}(H)} > 1\}$ be the open set in $\widehat{Q_0^{(\mathfrak{a})}}$ determined by $q \in Q_0^{(\mathfrak{a})}$ [**15**, Proposition 3.3.2]. Let us first show that Φ is continuous at each point $[\pi_\mathfrak{m}] \in \widehat{Q_0^{(\mathfrak{a})}}$. So, let $\mathcal{U}_0 \in \mathfrak{U}(\Phi([\pi_\mathfrak{m}]))$ be arbitrary. We are going to find $q_\mathfrak{m} \in Q_0^{(\mathfrak{a})}$ with $\mathfrak{m} \in \mathfrak{V}_{q_\mathfrak{m}}$ and $\Phi(\mathfrak{V}_{q_\mathfrak{m}}) \subseteq \mathcal{U}_0$.

For $\mathfrak{m} = \zeta \in {}^0S^*X$ with $\pi(\zeta) \notin \partial X$, i.e. $\zeta \in \mathcal{U}_0 = U \subseteq {}^0S^*X$, we choose $f \in \mathcal{C}_c^\infty(U)$ with $f(\zeta) = 2$, and let $q_\mathfrak{m} := \tau_0(A)$ where $A \in \Psi_0^0(X; {}^0\Omega^{1/2})$ satisfies ${}^0\sigma^{(0)}(A) = f$. For $\mathfrak{m} = \zeta \in {}^0S^*X|_{\partial X}$, let $A \in \mathcal{B}_0^{(\mathfrak{a})}(X, {}^0\Omega^{1/2})$ be as in Lemma 7.6.4. Then $q_\mathfrak{m} := \tau_0(A)$ has the desired properties.

In case $\mathfrak{m} = (q, \lambda) \in \partial X \times \Gamma_\mathfrak{a}$, let $q_\mathfrak{m} := \tau_0(A)$ where $A \in \mathcal{B}_0^{(\mathfrak{a})}(X, {}^0\Omega^{1/2})$ is as in Lemma 7.6.5. Finally, for $\mathfrak{m} = \widehat{\eta} \in S^*\partial X$, let $\mathcal{N} \in \mathcal{C}_c(W, \mathcal{K}(\varrho_0^\mathfrak{a} L^2(M, {}^{b,c}\Omega^{1/2})))$ be with $\|\mathcal{N}(\widehat{\eta})\|_{\mathcal{L}(\varrho_0^\mathfrak{a} L^2(M, {}^{b,c}\Omega^{1/2}))} > 1$. Then $(0, \mathcal{N}) \in Q_0^{(\mathfrak{a})}$ has the desired properties.

Therefore, $\Phi : \widehat{Q_0^{(\mathfrak{a})}} \to \mathfrak{M}$ is continuous. To show that $\Psi := \Phi^{-1} : \mathfrak{M} \to \widehat{Q_0^{(\mathfrak{a})}}$ is continuous as well, let $\mathfrak{m} \in \mathfrak{M}$ be arbitrary, and choose $q_\mathfrak{m} = \tau_0(A)$ for some $A \in \Psi_0^0(X; {}^0\Omega^{1/2})$ with $\Psi(\mathfrak{m}) \in \mathfrak{V}_{q_\mathfrak{m}}$. Recall that by [**15**, Lemma 3.3.3] the open sets $\mathfrak{V}_q$, $q = \tau_0(A)$, $A \in \Psi_0^0(X; {}^0\Omega^{1/2})$, form a local base for the Jacobson topology on $\widehat{Q_0^{(\mathfrak{a})}}$. We are going to find $\mathcal{U}_0 \in \mathfrak{U}(\mathfrak{m})$ with $\Psi(\mathcal{U}_0) \subseteq \mathfrak{V}(q_\mathfrak{m})$, i.e. $\|\pi_{\mathfrak{m}'}(\tau_0(A))\| > 1$ for all $\mathfrak{m}' \in \mathcal{U}_0$.

Let $\mathfrak{m} = \zeta_0 \in {}^0S^*X$ be with $\pi(\zeta_0) \notin \partial X$ and $|{}^0\sigma^{(0)}(A)(\zeta_0)| > 1$. By the continuity of ${}^0\sigma^{(0)}(A)$ there exists an open neighborhood $U \subseteq {}^0S^*X$ of ζ with $U \cap {}^0S^*X|_{\partial X} = \emptyset$ and $|{}^0\sigma^{(0)}(A)(\zeta)| > 1$ for all $\zeta \in U$. For $\zeta_0 \in {}^0S^*X|_{\partial X}$, let $j_{\varrho_N}(\eta_0, (q_0, \xi_0)) \in {}^0T^*X \setminus \{0\}$ be a lift of ζ_0. Then we have $|{}^0\sigma^{(0)}(A)(\zeta_0)| > 1$ and by the continuity of ${}^0\sigma^{(0)}(A)$ there exists an open neighbourhood $U \subseteq {}^0S^*X$ of ζ_0 with $|{}^0\sigma^{(0)}(A)(\zeta)| > 1$ for all $\zeta \in U$.

In case $\eta_0 \in T^*\partial X \setminus \{0\}$, we obtain by the compatibility condition (7.6)

$$1 < |{}^0\sigma^{(0)}(A)(\zeta_0)| = \left|I_c\left(\mathcal{N}_{\varrho_N}^\nu(A)(\widehat{\eta}_0)\right)(\xi_0/|\eta_0|_s)\right| \leq$$
$$\leq \|I_c\left(\mathcal{N}_{\varrho_N}^\nu(\widehat{\eta}_0)\right)\|_{\mathcal{B}(;\mathbb{R})} \leq \|\mathcal{N}_{\varrho_N}^\nu(A)(\widehat{\eta}_0)\|_{\mathcal{L}(\varrho_0^\mathfrak{a} L^2(M, {}^{b,c}\Omega^{1/2}))}$$

because $*$-morphisms between C^*-algebras are always norm-decreasing. Thus, there exists a neighborhood $W \subseteq S^*\partial X$ of $\widehat{\eta}_0$ with $\|\mathcal{N}_{\varrho_N}^\nu(A)(\widehat{\eta})\|_{\mathcal{L}(\varrho_0^\mathfrak{a} L^2(M, {}^{b,c}\Omega^{1/2}))} > 1$ for all $\widehat{\eta} \in W$, and $\mathcal{U}_0 := U \uplus W$ has the desired properties.

If $\eta_0 = 0$, the compatibility condition (7.5) yields for all $\widehat{\eta} \in \pi^{-1}(q_0) \subseteq S^*\partial X$

$$1 < |{}^0\sigma^{(0)}(A)(\zeta_0)| = \left|{}^{b,c}\sigma^{(0,0)}(\mathcal{N}_{\varrho_N}^\nu(A)(\widehat{\eta}))(z, \xi_0)\right| \leq \|\mathcal{N}_{\varrho_N}^\nu(A)(\widehat{\eta})\|_{\mathcal{L}(\varrho_0^\mathfrak{a} L^2(M, {}^{b,c}\Omega^{1/2}))},$$

thus, there exists an open neighborhood $V_1 \subseteq \partial X$ of q_0 such that for all $\widehat{\eta} \in \pi^{-1}(V_1)$ we have $\|\mathcal{N}_{\varrho_N}^\nu(A)(\widehat{\eta})\|_{\mathcal{L}(\varrho_0^\mathfrak{a} L^2(M, {}^{b,c}\Omega^{1/2}))} > 1$. Moreover, because of $\eta_0 = 0$ we have $\xi_0 \neq 0$, and the compatibility condition (7.8) gives for all $\widehat{\eta} \in S^*\partial X$ with $\pi(\widehat{\eta}) = q_0$

$$\left|\sigma_\mathcal{B}^{(0)}\left(I_b^{(\mathfrak{a})}\left(\mathcal{N}_{\varrho_N}^\nu(A)(\widehat{\eta})\right)\right)(\xi_0)\right| = \left|{}^{b,c}\sigma^{(0,0)}(\mathcal{N}_{\varrho_N}^\nu(A)(\widehat{\eta}))(0, \xi_0)\right| = |{}^0\sigma^{(0)}(A)(\zeta_0)| > 1.$$

By the exactness of (3.13), there exists $R > 0$ such that for all $\widehat{\eta} \in \pi^{-1}(q_0) \subseteq S^*\partial X$ and all $\xi \geq R$ we have

$$\left| I_b^{(\mathfrak{a})} \left(\mathcal{N}_{\varrho_N}^\nu(A)(\widehat{\eta}) \right) (\xi \operatorname{sign} \xi_0 - i\mathfrak{a}) \right| > \frac{1 + |{}^0\sigma^{(0)}(A)(\zeta_0)|}{2} > 1,$$

thus, we find an open neighborhood $V_2 \subseteq \partial X$ of q_0, and an open conical subset $\mathcal{W}_c$ of $\Gamma_\mathfrak{a}$ with $t\xi_0 - i\mathfrak{a} \in \mathcal{W}_c$ and $|I_b^{(\mathfrak{a})} \left(\mathcal{N}_{\varrho_N}^\nu(A)(\widehat{\eta}) \right)(\lambda)| > 1$ for all $(\widehat{\eta}, \lambda) \in \pi^{-1}(V_2) \times \mathcal{W}_c$. Then $\mathcal{U}_0 := U \uplus (V_1 \cap V_2 \times \mathcal{W}_c) \uplus \pi^{-1}(V_1 \cap V_2)$ has the desired properties.

If $\mathfrak{m} = (q_0, \lambda_0) \in \partial X \times \Gamma_\mathfrak{a}$, then we have for all $\widehat{\eta} \in \pi^{-1}(q_0) \subseteq S^*\partial X$

$$\begin{aligned} 1 < \left| I_b^{(\mathfrak{a})} \left(\mathcal{N}_{\varrho_N}^\nu(A)(\widehat{\eta}) \right)(\lambda_0) \right| &\leq \| I_b^{(\mathfrak{a})} \left(\mathcal{N}_{\varrho_N}^\nu(A)(\widehat{\eta}) \right) \|_{\mathcal{B}(;\Gamma_\mathfrak{a})} \\ &\leq \| \mathcal{N}_{\varrho_N}^\nu(A)(\widehat{\eta}) \|_{\mathcal{L}(\varrho_0^\mathfrak{a} L^2(M, {}^{b,c}\Omega^{1/2}))}, \end{aligned}$$

thus, there exists an open neighborhood $V \subseteq \partial X$ of q_0, and an open neighborhood $W \subseteq \Gamma_\mathfrak{a}$ of λ_0 such that we have $\left| I_b^{(\mathfrak{a})} \left(\mathcal{N}_{\varrho_N}^\nu(A)(\widehat{\eta}) \right)(\lambda) \right| > 1$ for all $(\widehat{\eta}, \lambda) \in \pi^{-1}(V) \times W$, and $\| \mathcal{N}_{\varrho_N}^\nu(A)(\widehat{\eta}) \|_{\mathcal{L}(\varrho_0^\mathfrak{a} L^2(M, {}^{b,c}\Omega^{1/2}))} > 1$ for all $\widehat{\eta} \in \pi^{-1}(V)$, and we can take $\mathcal{U}_0 := (V \times W) \uplus \pi^{-1}(V)$. Finally, for $\mathfrak{m} = \widehat{\eta} \in S^*\partial X$, we have $\| \mathcal{N}_{\varrho_N}^\nu(A)(\widehat{\eta}) \|_{\mathcal{L}(\varrho_0^\mathfrak{a} L^2(M, {}^{b,c}\Omega^{1/2}))} > 1$ and we find a neighborhood $\mathcal{U}_0 := W \subseteq S^*\partial X$ of $\widehat{\eta}$ as desired by the continuity of $\mathcal{N}_{\varrho_N}^\nu(A)$. □

REMARK 7.6.7. As observed in the proof of Proposition 7.5.2, the C^*-algebra $\mathcal{B}_0^{(\mathfrak{a})}(X, {}^0\Omega^{1/2})/\mathcal{I}_1 \cong Q_0^{(\mathfrak{a})}/J_1$ is commutative, hence it is isomorphic to the C^*-algebra of all continuous functions on a compact Hausdorff space $\mathcal{M}$, the space of all *maximal ideals* equipped with the Gelfand topology. By [15, Proposition 3.2.1], the space $\mathcal{M}$ is isomorphic to ${}^0S^*X \uplus (\partial X \times \Gamma_\mathfrak{a})$ where the compact Hausdorff topology on the latter is given by the Jacobson topology on $\mathfrak{M}$ and the embedding ${}^0S^*X \uplus (\partial X \times \Gamma_\mathfrak{a}) \hookrightarrow \mathfrak{M}$.

CHAPTER 8

Ψ^*-algebras of 0-pseudodifferential operators

8.1. Submultiplicative Ψ^*-algebras

We are now going to consider the 0-calculus from the point of view of topological algebras. Let us first recall the following notions.

DEFINITION 8.1.1. Let $\mathcal{A}$ be a complete, locally convex Fréchet-algebra with unit e.
(a) $\mathcal{A}$ is called *submultiplicative* provided there exists a family $(p_j)_{j \in \mathbb{N}}$ of seminorms that generates the topology of $\mathcal{A}$, satisfies $p_j(e) = 1$ and $p_j(ab) \leq p_j(a)p_j(b)$ for all $a, b \in \mathcal{A}$ and all $j \in \mathbb{N}$.
(b) If there exists a continuous inclusion $i : \mathcal{A} \hookrightarrow \mathcal{B}$ of $\mathcal{A}$ to a unital C^*-algebra $\mathcal{B}$ such that $i(\mathcal{A})$ is a symmetric subalgebra of $\mathcal{B}$, $i(e)$ is the unit element of $\mathcal{B}$, and

(8.1) $$i^{-1}(\mathcal{B}^{-1}) = \mathcal{A}^{-1}$$

holds for the groups of invertible elements $\mathcal{A}^{-1}$ resp. $\mathcal{B}^{-1}$ of $\mathcal{A}$ resp. $\mathcal{B}$, then $i(\mathcal{A})$ is called a Ψ^*-*algebra in* $\mathcal{B}$. The inclusion i is often omitted in the notation.

Note that submultiplicative Fréchet-algebras can be realized as the projective limit of a sequence of Banach algebras. The notion of Ψ^*-algebras has been introduced by Gramsch [23, Definition 5.1], first announced in [22]. It is an immediate consequence of the crucial property (8.1) of *spectral invariance* that the group of invertible elements of a Ψ^*-algebra is open in the Fréchet topology of $\mathcal{A}$.

Let us give some important examples of submultiplicative Ψ^*-algebras.

EXAMPLE 8.1.2. (a) Let $\mathcal{A}$ be a submultiplicative Ψ^*-algebra in the C^*-algebra $\mathcal{B}$ and P be a smooth, compact manifold with or without boundary. Then $\mathcal{C}^\infty(P, \mathcal{A})$ is a submultiplicative Ψ^*-algebra in the C^*-algebra $\mathcal{C}(P, \mathcal{B})$.
(b) If $\varphi : \mathcal{B}_0 \to \mathcal{B}_1$ is a $*$-morphism of C^*-algebras and $\mathcal{A}_1 \subseteq \mathcal{B}_1$ is a submultiplicative Ψ^*-algebra, then $\mathcal{A}_0 := \varphi^{-1}(\mathcal{A}_1)$ is a submultiplicative Ψ^*-algebra in $\mathcal{B}_0$ – see [40, Lemma 2.6] and [98, Lemma 3.2].
(c) If $(\mathcal{A}_j)_{j \in \mathbb{N}}$ is a sequence of (submultiplicative) Ψ^*-algebras in a C^*-algebra $\mathcal{B}$, then the intersection $\bigcap_{j \in \mathbb{N}} \mathcal{A}_j$ is a (submultiplicative) Ψ^*-algebra in $\mathcal{B}$ as well.
(d) The algebra of classical symbols $S^0_{cl}(;\mathbb{R})$ with pointwise product is a submultiplicative Ψ^*-algebra in the C^*-algebra $\mathcal{C}_b(\mathbb{R})$ of bounded continuous functions on the line. This follows for instance from (a) and the fact that $S^0_{cl}(;\mathbb{R})$ corresponds under radial compactification to the space of smooth functions on the half-circle.

(e) This example is a special case of a general construction in [**30**], see also [**40**]. Let H be a Hilbert space, and $V : H \supseteq \mathcal{D}(V) \to H$ be a closed, densely defined symmetric operator. Furthermore, let $\mathcal{B}(V)$ be the space of all $a \in \mathcal{L}(H)$ such that $a\mathcal{D}(V) \subseteq \mathcal{D}(V)$ and $Va - aV : \mathcal{D}(V) \to H$ extends to a bounded operator $\delta_V(a) \in \mathcal{L}(H)$. Then

$$i\delta_V : \mathcal{L}(H) \supseteq \mathcal{D}(i\delta_V) := \{a \in \mathcal{B}(V) : a^* \in \mathcal{B}(V)\} \longrightarrow \mathcal{L}(H) : a \longmapsto i\delta_V(a)$$

is a symmetric, closed $*$-derivation, however, not necessarily densely defined. Further, let $\Psi_1^V := \mathcal{D}(V)$, and $\Psi_{n+1}^V := \{a \in \Psi_n^V : i\delta_V(a) \in \Psi_n^V\}$, $n \in \mathbb{N}$, be equipped with the iterated graph norms. Then

$$\Psi^V := \bigcap_{n=1}^{\infty} \Psi_n^V \hookrightarrow \mathcal{L}(H)$$

equipped with the projective limit topology is a submultiplicative Ψ^*-algebra in $\mathcal{L}(H)$.

8.2. Ψ^*-completions of b-c- and 0-calculus

It is well-known that the small b-calculus on a compact manifold with boundary does no admit a topology making the algebra of operators of order 0 to a topological algebra with an open group of invertible elements [**40**, Theorem 3.4]. Being similar to the b-calculus, it is not surprising that the same is true for the 0- and the b-c-calculus as well. Since the proof of this result is quite similar to the proof of [**40**, Theorem 3.4], we sketch the argument only.

PROPOSITION 8.2.1. *Neither on the algebra $\Psi_0^0(X; {}^0\Omega^{1/2})$ nor on the algebra $\Psi_{b,c}^0(M; {}^{b,c}\Omega^{1/2})$ there exists a topology giving $\Psi_0^0(X; {}^0\Omega^{1/2})$ resp. $\Psi_{b,c}^0(M; {}^{b,c}\Omega^{1/2})$ the structure of a topological algebra with an open group of invertible elements. In particular, neither $\Psi_0^0(X; {}^0\Omega^{1/2})$ nor $\Psi_{b,c}^0(M; {}^{b,c}\Omega^{1/2})$ can be realized as a Ψ^*-algebra.*

For simplicity, let us abbreviate $\Psi := \Psi_0^0(X; {}^0\Omega^{1/2})$ or $\Psi := \Psi_{b,c}^0(M; {}^{b,c}\Omega^{1/2})$.
Proof: Let $h : \mathbb{C} \to \mathbb{C} : z \mapsto e^{-z^2}$. Then it is not hard to see that there exists $A \in \Psi$ of order $-\infty$ with $I_A = h$ resp. $I_b(A) = h$. Suppose that there exists a topology on Ψ making Ψ to a topological algebra with an open group of invertible elements, and consider the continuous path $\gamma : [0,1] \to \Psi : t \mapsto \mathrm{id} + tA$. Because of $\gamma(0) = \mathrm{id}$ there exists $t_0 > 0$ and $\beta(t) \in \Psi$ such that $\beta(t)\gamma(t) = \gamma(t)\beta(t) = \mathrm{id}$ for all $0 \leq t < t_0$. Therefore,

$$1 = I_{\beta(t)}(z)(q)(1 + th(z)) \text{ resp. } 1 = I_b(\beta(t))(z)(1 + th(z))$$

for all $z \in \mathbb{C}$ which is a contradiction because of $h(\mathbb{C}) = \mathbb{C} \setminus \{0\}$. $\square$

Let us also note that to the best of our knowledge it is not known whether there exists a submultiplicative Fréchet-topology on $\Psi_0^0(X; {}^0\Omega^{1/2})$ or $\Psi_{b,c}^0(M; {}^{b,c}\Omega^{1/2})$.

Since $\mathcal{B} := \mathcal{B}_0^{(a)}(X, {}^0\Omega^{1/2})$ and $\mathcal{B} := \mathcal{B}_{b,c}^{(a_0,a_1)}(M, {}^{b,c}\Omega^{1/2})$ are certainly submultiplicative Ψ^*-algebras (even C^*-algebras) that contain the algebras $\Psi_0^0(X; {}^0\Omega^{1/2})$ and $\Psi_{b,c}^0(M; {}^{b,c}\Omega^{1/2})$ as dense subalgebras it is tempting to ask whether there exists interesting, submultiplicative Ψ^*-algebra $\mathcal{A}$ between Ψ and $\mathcal{B}$ that cover additionally (at least some of) the $\mathcal{C}^\infty$-properties of the algebra Ψ that are completely invisible for the C^*-algebra $\mathcal{B}$. This approach has been suggested by Gramsch.

8.2. Ψ^*-COMPLETIONS OF b-c- AND 0-CALCULUS

The methods for constructing those Ψ^*-*completions* were developed in great generality in [30] and [40]; they were used in [41] to construct a Ψ^* completion of the small b-calculus on a compact manifold with boundary. Since the technicalities are almost identical, we only state the results and give a hint how to prove them. First, we need the following notion.

DEFINITION 8.2.2. Let H^s, $s \in \mathbb{N}_0$ be a *scale of Hilbert spaces*, i.e. for all $s \in \mathbb{N}_0$ we have $H^{s+1} \subseteq H^s$ and the inclusion $H^{s+1} \hookrightarrow H^s$ is continuous. We say that *elliptic regularity with respect to the scale* $(H^s)_{s \in \mathbb{N}_0}$ holds for an operator $A \in \mathcal{L}(H^0)$ if $u \in H^s$ whenever $u \in H^0$ satisfies $Au \in H^s$.

In the context of Ψ^*-algebras the question of elliptic regularity has been studied in [28], see also [40, Theorem 2.11]. We start by considering Ψ^*-completions of the b-c-calculus on M.

THEOREM 8.2.3. *For any* $\mathfrak{a}_0, \mathfrak{a}_1 \in \mathbb{R}$, *there exists a submultiplicative* Ψ^*-*algebra* $\mathcal{A}_{b,c}^{(\mathfrak{a}_0,\mathfrak{a}_1)}(M, {}^{b,c}\Omega^{1/2})$ *in the* C^*-*algebra* $\mathcal{B}_{b,c}^{(\mathfrak{a}_0,\mathfrak{a}_1)}(M, {}^{b,c}\Omega^{1/2})$ *such that*

(a) $\Psi_{b,c}^0(M; {}^{b,c}\Omega^{1/2})$ *is a dense subalgebra of* $\mathcal{B}_{b,c}^{(\mathfrak{a}_0,\mathfrak{a}_1)}(M, {}^{b,c}\Omega^{1/2})$.

(b) *Any* $A \in \mathcal{A}_{b,c}^{(\mathfrak{a}_0,\mathfrak{a}_1)}(M, {}^{b,c}\Omega^{1/2})$ *extends from* $\dot{\mathcal{C}}^\infty(M, {}^{b,c}\Omega^{1/2})$ *to a bounded operator*
$$A : \varrho_0^{\mathfrak{a}_0} \varrho_1^{\mathfrak{a}_1} H_{b,c}^s(M, {}^{b,c}\Omega^{1/2}) \to \varrho_0^{\mathfrak{a}_0} \varrho_1^{\mathfrak{a}_1} H_{b,c}^s(M, {}^{b,c}\Omega^{1/2}) \text{ for all } s \in \mathbb{N}_0\,.$$

(c) *If a Fredholm operator* $A : \varrho_0^{\mathfrak{a}_0} \varrho_1^{\mathfrak{a}_1} L^2(M, {}^{b,c}\Omega^{1/2}) \to \varrho_0^{\mathfrak{a}_0} \varrho_1^{\mathfrak{a}_1} L^2(M, {}^{b,c}\Omega^{1/2})$ *satisfies* $A \in \mathcal{A}_{b,c}^{(\mathfrak{a}_0,\mathfrak{a}_1)}(M, {}^{b,c}\Omega^{1/2})$, *then elliptic regularity holds for* A *with respect to the scale* $(\varrho_0^{\mathfrak{a}_0} \varrho_1^{\mathfrak{a}_1} H_{b,c}^s(M, {}^{b,c}\Omega^{1/2}))_{s \in \mathbb{N}_0}$.

(d) *Any* $A \in \mathcal{A}_{b,c}^{(\mathfrak{a}_0,\mathfrak{a}_1)}(M, {}^{b,c}\Omega^{1/2})$ *has smooth symbols, i.e. we have*
- ${}^{b,c}\sigma^{(0,0)}(A) \in \mathcal{C}^\infty({}^{b,c}S^*M)$,
- $I_c(A) \in S_{cl}^0(;\mathbb{R}_\xi)$, and
- $\left[\xi \mapsto I_b^{(\mathfrak{a}_0,\mathfrak{a}_1)}(A)(\xi - i\mathfrak{a}_0)\right] \in S_{cl}^0(;\mathbb{R}_\xi)$.

(e) *Let* $\omega_1, \omega_2 \in \mathcal{C}_c^\infty((0,1))$ *and* $A \in \mathcal{A}_{b,c}^{(\mathfrak{a}_0,\mathfrak{a}_1)}(M, {}^{b,c}\Omega^{1/2})$. *Then* $\omega_1 A \omega_2$ *is an ordinary, compactly supported pseudodifferential operator on the open manifold* $(0,1)$.

Proof: The construction of the Ψ^*-completion $\mathcal{A}_{b,c}^{(\mathfrak{a}_0,\mathfrak{a}_1)}(M, {}^{b,c}\Omega^{1/2})$ requires several steps. First, using Example 8.1.2(b) to pull-back the submultiplicative Ψ^*-algebras $S_{cl}^0(;\mathbb{R})$ and $\mathcal{C}^\infty({}^{b,c}S^*M)$, we obtain a submultiplicative Ψ^*-algebra with (d). An extension of Example 8.1.2 to (countable) many closed operators induced by b-c-vector fields, yields (b) and (e). Note that $\Psi_{b,c}^0(M; {}^{b,c}\Omega^{1/2})$ is stable under commutators with b-c-vector fields by Theorem 3.5.2; for (e) recall that pseudodifferential operators can be characterized by the boundedness of iterated commutators with vector fields [11, 41]. Moreover, elliptic regularity (c) holds by the general results of Gramsch and Kalb in [28]. Finally, by taking the closure of $\Psi_{b,c}^0(M; {}^{b,c}\Omega^{1/2})$ within the intersection of the submultiplicative Ψ^*-algebras obtained above leads to a submultiplicative Ψ^*-algebra $\mathcal{A}_{b,c}^{(\mathfrak{a}_0,\mathfrak{a}_1)}(M, {}^{b,c}\Omega^{1/2})$ as desired. $\square$

We are now going to present the main result of this section.

THEOREM 8.2.4. *For any* $\mathfrak{a} \in \mathbb{R}$, *there exists a submultiplicative* Ψ^*-*algebra* $\mathcal{A}_0^{(\mathfrak{a})}(X, {}^0\Omega^{1/2})$ *in the* C^*-*algebra* $\mathcal{B}_0^{(\mathfrak{a})}(X, {}^0\Omega^{1/2})$, *such that*

(a) $\Psi_0^0(X; {}^0\Omega^{1/2})$ is a dense subalgebra.
(b) Any $A \in \mathcal{A}_0^{(\mathfrak{a})}(X, {}^0\Omega^{1/2})$ extends from $\dot{\mathcal{C}}^\infty(X, {}^0\Omega^{1/2})$ to a bounded operator
$$A : \varrho_N^\mathfrak{a} H_0^s(X, {}^0\Omega^{1/2}) \to \varrho_N^\mathfrak{a} H_0^s(X, {}^0\Omega^{1/2}) \ , \ s \in \mathbb{N}_0 \ .$$
Moreover, the assiociated bilinear map
$$\mathcal{A}_0^{(\mathfrak{a})}(X, {}^0\Omega^{1/2}) \times \varrho_N^\mathfrak{a} H_0^s(X, {}^0\Omega^{1/2}) \to \varrho_N^\mathfrak{a} H_0^s(X, {}^0\Omega^{1/2})$$
is jointly continuous.
(c) If a Fredholm operator $A : \varrho_N^\mathfrak{a} L^2(X, {}^0\Omega^{1/2}) \to \varrho_N^\mathfrak{a} L^2(X, {}^0\Omega^{1/2})$ satisfies $A \in \mathcal{A}_0^{(\mathfrak{a})}(X, {}^0\Omega^{1/2})$, then elliptic regularity holds for A with respect to the scale $(\varrho_N^\mathfrak{a} H_0^s(X, {}^0\Omega^{1/2}))_{s \in \mathbb{N}_0}$.
(d) Any $A \in \mathcal{A}_0^{(\mathfrak{a})}(X, {}^0\Omega^{1/2})$ has smooth symbols, i.e.
- ${}^0\sigma^{(0)}(A) \in \mathcal{C}^\infty({}^0S^*X)$,
- $\mathcal{N}_{\varrho_N}^\nu(A) \in \mathcal{C}^\infty(S^*\partial X, \mathcal{A}_{b,c}^{(\mathfrak{a},0)}(M, {}^{b,c}\Omega^{1/2})) \cap \mathcal{B}_{b,c}^{(\mathfrak{a})}(M, {}^{b,c}\Omega^{1/2}; S^*\partial X)$,
where $\mathcal{A}_{b,c}^{(\mathfrak{a},0)}(M, {}^{b,c}\Omega^{1/2})$ is the submultiplicative Ψ^*-algebra provided by Theorem 8.2.3.
(e) Let $\omega_1, \omega_2 \in \mathcal{C}_c^\infty(\operatorname{int} X)$ and $A \in \mathcal{A}_0^{(\mathfrak{a})}(X, {}^0\Omega^{1/2})$. Then $\omega_1 A \omega_2$ is an ordinary, compactly supported pseudodifferential operator in the interior $\operatorname{int} X$ of X.

Proof: The main steps in the proof are mainly identical to the proof of Theorem 8.2.3, hence we do not repeat them. For
$$\mathcal{N}_{\varrho_N}^\nu(A) \in \mathcal{C}^\infty(S^*\partial X, \mathcal{A}_{b,c}^{(\mathfrak{a},0)}(M, {}^{b,c}\Omega^{1/2})) \cap \mathcal{B}_{b,c}^{(\mathfrak{a})}(M, {}^{b,c}\Omega^{1/2}; S^*\partial X)$$
note that the algebra on the right hand side is a submultiplicative Ψ^*-algebra in the C^*-algebra $\mathcal{B}_{b,c}^{(\mathfrak{a})}(M, {}^{b,c}\Omega^{1/2}; S^*\partial X)$ by Example 8.1.2(b). □

REMARK 8.2.5. (a) Since the cusp-calculus on a compact manifold with boundary is known to be closed under holomorphic functional calculus, we can also find a Ψ^*-algebra $\Psi_{b,c}^0(M; {}^{b,c}\Omega^{1/2}) \subseteq \widetilde{\mathcal{A}} \subseteq \mathcal{B}_{b,c}^{(\mathfrak{a}_0,\mathfrak{a}_1)}(M, {}^{b,c}\Omega^{1/2})$ such that for $\omega_0, \omega_1 \in \mathcal{C}_c^\infty((0,1])$ and $A \in \widetilde{\mathcal{A}}$, the operator $\omega_0 A \omega_1$ is a cusp operator near $1 \in M$. However, we do not yet know whether there exists also a submultiplicative Fréchet-topology on $\widetilde{\mathcal{A}}$.
(b) The submultiplicative Ψ^*-algebras $\mathcal{A}_{b,c}^{(\mathfrak{a},0)}(M, {}^{b,c}\Omega^{1/2})$ resp. $\mathcal{A}_0^{(\mathfrak{a})}(X, {}^0\Omega^{1/2})$ depend in fact on various choices, for instance sequences of b-c- resp. 0-vector fields on M resp. X. However, by considering the intersection over all possible choices we obtain a complete, topological algebra $\mathcal{A}$ with jointly continuous multiplication and continuous inversion that still enjoys properties (a)–(e) of Theorem 8.2.3 resp. Theorem 8.2.4, and satisfies the spectral invariance condition (8.1); in particular, $\mathcal{A}$ is closed under functional calculus. Note that $\mathcal{A}$ is in general not Fréchet anymore.

APPENDIX A

Spaces of conormal functions

We are going to discuss the relation between certain spaces of smooth functions $f : \mathbb{R}_z^N \setminus \{0\} \to \mathbb{C}$ with different regularity as $z \to 0$. For simplicity, we always assume $f(z) = 0$ for all z with $|z| \geq R$ for some $R > 0$. Let us denote by

$$\mathcal{V}_E(\mathbb{R}^N; \{0\}) := \{V \in \mathcal{C}^\infty(\mathbb{R}^N, T\mathbb{R}^N) : V(0) = 0\}$$

the space of all smooth vector fields on $\mathbb{R}^N$ vanishing at the origin. It is well-known that the *linear vector fields*

$$\mathcal{M}_{\mathrm{lin}}(\mathbb{R}^N) := \{z_j \partial_{z_k} \in \mathcal{V}_E(\mathbb{R}^N; \{0\}) : j, k = 1, \ldots, N\}$$

form a generating set of $\mathcal{V}_E(\mathbb{R}^N; \{0\})$ over $\mathcal{C}^\infty(\mathbb{R}^N)$.

DEFINITION A.1. Let $s \in \mathbb{R}$ and $R > 0$ be arbitrary. Then we define

$$\mathcal{C}_R^\infty(\mathbb{R}^N; \{0\}) := \{f \in \mathscr{S}'(\mathbb{R}^N) : \mathrm{sing\,supp}\, f \subseteq \{0\} \text{ and } f(z) = 0 \text{ for all } |z| \geq R\},$$

$$IH_R^s(\mathbb{R}^N; \{0\}) := \{f \in \mathcal{C}_R^\infty(\mathbb{R}^N; \{0\}) : \mathcal{V}_E(\mathbb{R}^N; \{0\})^\ell f \subseteq H^s(\mathbb{R}^N) \text{ for all } \ell \in \mathbb{N}_0\},$$

$$IL_R^2(\mathbb{R}^N; \{0\}) := \{f \in \mathcal{C}_R^\infty(\mathbb{R}^N; \{0\}) : \mathcal{V}_E(\mathbb{R}^N; \{0\})^\ell f \subseteq L^2(\mathbb{R}^N) \text{ for all } \ell \in \mathbb{N}_0\},$$

$$IL_R^\infty(\mathbb{R}^N; \{0\}) := \{f \in \mathcal{C}_R^\infty(\mathbb{R}^N; \{0\}) : \mathcal{V}_E(\mathbb{R}^N; \{0\})^\ell f \subseteq L^\infty(\mathbb{R}^N) \text{ for all } \ell \in \mathbb{N}_0\}.$$

REMARK A.2. Since $\mathcal{V}_E(\mathbb{R}^N; \{0\})$ is a finite-dimensional $\mathcal{C}^\infty(\mathbb{R}^N)$-module, and everything happens in the ball $\{|z| \leq R\}$, we can certainly replace the space $\mathcal{V}_E(\mathbb{R}^N; \{0\})$ by the set $\mathcal{M}_{\mathrm{lin}}(\mathbb{R}^N)$ of linear vector fields in the definition above. In particular, we obtain natural Fréchet topologies on the spaces $IH_R^s(\mathbb{R}^N; \{0\})$, $IL_R^2(\mathbb{R}^N; \{0\})$, and $IL_R^\infty(\mathbb{R}^N; \{0\})$. Consequently, we can endow the inductive limits

$$IH_c^s(\mathbb{R}^N; \{0\}) := \bigcup_{R>0} IH_R^s(\mathbb{R}^N; \{0\}),$$

$$IL_c^2(\mathbb{R}^N; \{0\}) := \bigcup_{R>0} IL_R^2(\mathbb{R}^N; \{0\}), \text{ and}$$

$$IL_c^\infty(\mathbb{R}^N; \{0\}) := \bigcup_{R>0} IL_R^\infty(\mathbb{R}^N; \{0\})$$

with the corresponding inductive limit topologies

Let $a \in S^m(; \mathbb{R}_\zeta^N)$ be a symbol of order m, i.e. a smooth function $a : \mathbb{R}_\zeta^N \to \mathbb{C}$ satisfying

(A.1) $$|\partial_\zeta^\alpha a(\zeta)| \leq \mathrm{const}\,(\alpha, a) <\zeta>^{m-|\alpha|}$$

for all $\zeta \in \mathbb{R}_\zeta^N$. The inverse Fourier transform $\mathcal{F}_{\zeta \to z}^{-1} a$ of such a symbol is known to be smooth in $\mathbb{R}_z^N \setminus \{0\}$ and rapidly decreasing as $|z| \to \infty$. Moreover, it is compactly supported in $\{|z| \leq R\}$ provided a extends to an entire function $\mathbb{C}^N \to \mathbb{C}$ with additional locally uniform estimates by the Paley-Wiener theorem.

DEFINITION A.3. For $R > 0$ and $m \in \mathbb{R}$, let $I_R^m(\mathbb{R}^N; \{0\})$ be the space of all $f = \mathcal{F}_{\zeta \to z}^{-1} a|_{\mathbb{R}_z^N \setminus \{0\}}$ with $f(z) = 0$ for $|z| \geq R$ and $a \in S^{m-\frac{N}{4}}(;\mathbb{R}_\zeta^N)$. We denote by $I_c^m(\mathbb{R}^N; \{0\}) := \bigcup_{R>0} I_R^m(\mathbb{R}^N; \{0\})$ the corresponding inductive limit.

The best constants in (A.1), and the corresponding constants of the entire extension induce a Fréchet topology on $I_R^m(\mathbb{R}^N; \{0\})$, hence also a locally convex topology on $I_c^m(\mathbb{R}^N; \{0\})$. The canonical inclusions $I_R^m(\mathbb{R}^N; \{0\}) \hookrightarrow \mathcal{C}^\infty(\mathbb{R}_z^N \setminus \{0\})$ resp. $I_c^m(\mathbb{R}^N; \{0\}) \hookrightarrow \mathcal{C}^\infty(\mathbb{R}_z^N \setminus \{0\})$ are then continuous.

The next result can be proved, for instance, by a straightforward use of standard techniques for oscillatory integrals – for a similar, detailed computation see [**38**, Proposition 4.1.10].

LEMMA A.4. *Let* $\chi \in \mathcal{C}_c^\infty(\mathbb{R}^N)$ *be with* $\chi = 1$ *near* $z = 0$ *and* $\chi = 0$ *for* $|z| \geq R$. *Then the map*

$$\Psi_\chi : S^{m-\frac{N}{4}}(;\mathbb{R}_\zeta^N) \longrightarrow I_R^m(\mathbb{R}^N; \{0\}) : a \longmapsto \chi \mathcal{F}_{\zeta \to z}^{-1} a$$

is well-defined and continuous.

A proof for Proposition A.6 can be found for instance in [**63**, XV.30]. However, we repeat the essential steps for the sake of completeness. We start with a simple observation from elementary calculus.

LEMMA A.5. *Let* $f : \mathbb{R} \longrightarrow \mathbb{R}$ *be such that* $f(r) = 0$ *for* $r \leq 1$, *and assume there exists* $\gamma < -\frac{1}{2}$ *as well as* $g \in L^2(\mathbb{R})$ *with* $\partial_r f = r^\gamma g$. *Then we have* $f \in L^\infty(\mathbb{R})$.

PROPOSITION A.6. *Let* $s, m \in \mathbb{R}$ *and* $R > 0$ *be arbitrary.*
(a) $I_R^m(\mathbb{R}^N; \{0\}) \subseteq IH_R^s(\mathbb{R}^N; \{0\})$ *provided* $m < -(s + \frac{N}{4})$, *and*
(b) $IH_R^s(\mathbb{R}^N; \{0\}) \subseteq I_R^m(\mathbb{R}^N; \{0\})$ *provided* $m > -(s + \frac{N}{4})$.
The embeddings $I_R^m(\mathbb{R}^N; \{0\}) \hookrightarrow IH_R^s(\mathbb{R}^N; \{0\})$ *resp.* $IH_R^s(\mathbb{R}^N; \{0\}) \hookrightarrow I_R^m(\mathbb{R}^N; \{0\})$ *are then continuous.*

Proof: First note that the continuity of the canonical embeddings is a consequence of the closed graph theorem.

(a) Let $f \in I_R^m(\mathbb{R}^N; \{0\})$ be arbitrary. Using the set $\mathcal{M}_{\text{lin}}(\mathbb{R}^N)$ of linear vector fields we obtain

$f \in IH_R^s(\mathbb{R}^N; \{0\})$
$\iff V_1 \cdots V_\ell f \in H^s(\mathbb{R}_z^N), V_j \in \mathcal{M}_{\text{lin}}(\mathbb{R}^N), j = 1, \ldots, \ell, \ell \in \mathbb{N}_0$
$\iff <\zeta>^s \mathcal{F}_{z \to \zeta}(V_1 \cdots V_\ell f) \in L^2(\mathbb{R}_\zeta^N), V_j \in \mathcal{M}_{\text{lin}}(\mathbb{R}^N), j = 1, \ldots, \ell, \ell \in \mathbb{N}_0$
$\iff <\zeta>^s \zeta^\alpha \partial_\zeta^\beta \mathcal{F}_{z \to \zeta} f \in L^2(\mathbb{R}_\zeta^N)$ for all $\alpha, \beta \in \mathbb{N}_0^N$ with $|\alpha| = |\beta|$

Because of $\mathcal{F}_{z \to \zeta} f \in S^{m-\frac{N}{4}}(;\mathbb{R}_\zeta^N)$ we have

$$\left| <\zeta>^s \zeta^\alpha \partial_\zeta^\beta \mathcal{F}_{z \to \zeta} f \right|^2 \leq \text{const} <\zeta>^{2s+2m-\frac{N}{2}} \in L^1(\mathbb{R}_\zeta^N) \iff s + m < -\frac{N}{4}$$

which completes the proof of (a).

(b) Let $f \in IH_R^s(\mathbb{R}^N; \{0\})$ be arbitrary, and denote by $a = \mathcal{F}_{z \to \zeta} f$ its Fourier transform. Because of $f \in IH_R^s(\mathbb{R}^N; \{0\})$ we have

(A.2) $\quad <\zeta>^s \zeta^\alpha \partial_\zeta^\beta a(\zeta) \in L^2(\mathbb{R}_\zeta^N)$ for all $\alpha, \beta \in \mathbb{N}_0^N$ with $|\alpha| = |\beta|$.

Thus, for $a \in S^{m-\frac{N}{4}}(; \mathbb{R}_\zeta^N)$ it suffices to check that (A.2) implies

$$|a(\zeta)| \leq \mathrm{const}\,(a) <\zeta>^{m-\frac{N}{4}} \text{ for all } \zeta \in \mathbb{R}_\zeta^N$$

if $m + \frac{N}{4} > -s$. Let $\beta_p : \overline{\mathbb{R}}_+ \times S^{N-1} \longrightarrow \mathbb{R}_\zeta^N : (r, \omega) \longmapsto r\omega = \zeta$ be the usual polar coordinates. Since any differential operator of order k in $\mathbb{R}_\zeta^N$ lifts to a differential operators of the form

$$r^{-k} \sum_{j=0}^k P_j(r)(r\partial_r)^j \text{ with } P_j \in \mathcal{C}^\infty(\overline{\mathbb{R}}_+, \mathrm{Diff}^{k-j}(S^{N-1})),$$

(A.2) implies for all $k \in \mathbb{N}_0$ and all $P \in \mathrm{Diff}^*(S^{N-1})$

$$\partial_r^k P(\beta_p^* a) \in r^{-s-k} L^2(\overline{\mathbb{R}}_+ \times S^{N-1}; r^{N-1} dr\, d\omega),$$

where we used the invariance of the differential operators $\zeta^\alpha \partial_\zeta^\beta$ under the dilation $\zeta \mapsto t\zeta$ for $t > 0$ and $|\alpha| = |\beta|$. By elliptic regularity on S^{N-1} we obtain

(A.3) $\quad \left[r \longmapsto \sup_{\omega \in S^{N-1}} \left|\partial_r^k P(\beta_p^* a)(r, \omega)\right|\right] \in r^{-s-k} L^2(\overline{\mathbb{R}}_+; r^{N-1} dr) =$

$$= r^{-s-k-\frac{N-1}{2}} L^2(\overline{\mathbb{R}}_+; dr)$$

for all $k \in \mathbb{N}_0$ and all $P \in \mathrm{Diff}^*(S^{N-1})$. Since we are only interested in the behaviour as $r = |\zeta| \to \infty$, we can assume without loss of generality that $\beta_p^* a$ vanishes for $r \leq 1$. Thus, (A.3) implies in particular

$$\left[r \longmapsto \sup_{\omega \in S^{N-1}} \left|\partial_r r^q (\beta_p^* a)(r, \omega)\right|\right] \in r^\gamma L^2(\overline{\mathbb{R}}_+; dr)$$

provided $q, \gamma \in \mathbb{R}$ satisfy $q - 1 - s - \frac{N-1}{2} = \gamma$. Consequently, Lemma A.5 yields

$$|a(\zeta)| = |\beta_p^* a(|\zeta|, \frac{\zeta}{|\zeta|})| \leq \mathrm{const}\, |\zeta|^{-q}$$

for all q with $\gamma = q - 1 - s - \frac{N-1}{2} < -\frac{1}{2}$; in particular, it holds for $q = -m + \frac{N}{4} < s + \frac{N}{2}$ which completes the proof of (b). $\square$

DEFINITION A.7. Let $\gamma \in \mathbb{R}$ and $R > 0$ be arbitrary. Then we define

$$|z|^\gamma IL_R^2(\mathbb{R}^N; \{0\}) := \{f \in \mathscr{S}'(\mathbb{R}^N) : |z|^{-\gamma} f \in IL_R^2(\mathbb{R}^N; \{0\})\}, \text{ and}$$
$$|z|^\gamma IL_R^\infty(\mathbb{R}^N; \{0\}) := \{f \in \mathscr{S}'(\mathbb{R}^N) : |z|^{-\gamma} f \in IL_R^\infty(\mathbb{R}^N; \{0\})\}.$$

The inductive limit spaces $|z|^\gamma IL_c^2(\mathbb{R}^N; \{0\})$ and $|z|^\gamma IL_c^\infty(\mathbb{R}^N; \{0\})$ are defined accordingly.

Note that for $\frac{N}{2} < s$ elements in $IH_R^s(\mathbb{R}^N; \{0\})$ are continuous at $z = 0$ by Sobolev's Lemma. In fact they are Hölder continuous of exponent $\alpha < s - \frac{N}{2}$. Within our setting of compactly supported conormal functions this reads as follows. Let $\chi \in \mathcal{C}_c^\infty(\mathbb{R}_z^N)$ be with $\chi = 1$ near $z = 0$, and $\chi = 0$ for $|z| \geq R$. Then we have a map

$$T_\chi : IH_R^s(\mathbb{R}^N; \{0\}) \longrightarrow \mathcal{C}_R^\infty(\mathbb{R}^N; \{0\}) : f \longmapsto \chi(f - f(0)).$$

PROPOSITION A.8. *For each $0 < \gamma \leq 1$ and each $\frac{N}{2} + \gamma < s$ the map*
$$T_\chi : IH_R^s(\mathbb{R}^N; \{0\}) \longrightarrow |z|^\gamma IL_R^\infty(\mathbb{R}^N; \{0\})$$
is well-defined and continuous.

Proof: Note that it suffices to show $T_\chi(IH_R^s(\mathbb{R}^N; \{0\})) \subseteq |z|^\gamma IL_R^\infty(\mathbb{R}^N; \{0\})$. The continuity is then a consequence of the closed graph theorem.

Let $f \in IH_R^s(\mathbb{R}^N; \{0\})$ be arbitrary. We are going to show $|z|^{-\gamma} T_\chi f \in L^\infty(\mathbb{R}^N)$. Because of

(A.4) $\quad ||z|^{-\gamma}(T_\chi f)(z)|^2 \leq \underbrace{|z|^{-2\gamma} \int_{\mathbb{R}_\zeta^N} |e^{i\zeta z} - 1|^2 <\zeta>^{-2s} d\zeta}_{=: c_{\gamma s}(z) \geq 0} \|f\|_{H^s(\mathbb{R}^N)}^2,$

it remains to estimate $c_{\gamma s}(z)$ uniformly in z. This, however, is an easy consequence of $2(\gamma - s) < -N$, and $t^{-2\gamma}|e^{i\gamma} - 1|^2 \leq 4$ for all $t \in \mathbb{R} \setminus \{0\}$. Note that (A.4) follows first by Fourier inversion formula and Cauchy's inequality for $f \in \mathscr{S}(\mathbb{R}^N)$, and then by continuity for all $f \in H^s(\mathbb{R}^N)$.

Before proving $\mathcal{V}_E(\mathbb{R}^N; \{0\})^\ell (|z|^{-\gamma} T_\chi f) \subseteq L^\infty(\mathbb{R}^N)$, let us first note that for $V_j \in \mathcal{V}_E(\mathbb{R}^N; \{0\})$, $j = 1, \ldots, \ell$ we have

(A.5) $\quad\quad\quad\quad\quad\quad\quad V_\ell \cdots V_1 f|_{z=0} = 0.$

Indeed, choose $\varepsilon > 0$ with $s - 3\varepsilon > \frac{N}{2} + \gamma$. By Proposition A.6(b), $f \in I_R^m(\mathbb{R}^N; \{0\})$ for $m = \varepsilon - (s + \frac{N}{4})$, hence $\mathcal{F}_{z \to \zeta} f \in S^{\varepsilon - s - \frac{N}{2}}(;\mathbb{R}_\zeta^N)$ by Definition A.3. Since $\mathscr{S}(\mathbb{R}_\zeta^N) = S^{-\infty}(;\mathbb{R}_\zeta^N)$ is dense in $S^{\varepsilon - s - \frac{N}{2}}$ in the topology of $S^{2\varepsilon - s - \frac{N}{2}}(;\mathbb{R}_\zeta^N)$ [**35**, Proposition 1.1.11], we find a sequence $f_j \in \mathscr{S}(\mathbb{R}_z^N)$ with $\mathcal{F}_{z \to \zeta} f_j \to \mathcal{F}_{z \to \zeta} f$ in $S^{2\varepsilon - s - \frac{N}{2}}(;\mathbb{R}_\zeta^N)$, hence $\mathscr{S}(\mathbb{R}_z^N) \cap C_R^\infty(\mathbb{R}^N; \{0\}) \ni \Psi_\chi(\mathcal{F}_{z \to \zeta} f_j) = \chi f_j \to f$ in $I_R^{m'}(\mathbb{R}^N; \{0\})$ for $m' = 2\varepsilon - s - \frac{N}{4}$ by Lemma A.4, and, thus, by Proposition A.6(a) in $IH_R^{s'}(\mathbb{R}^N; \{0\})$ for $s' = -m' - \frac{N}{4} - \varepsilon = s - 3\varepsilon > \frac{N}{2} + \gamma$. Consequently, we have $V_\ell \cdots V_1 \chi f_j \to V_\ell \cdots V_1 f$ in $H^{s'}(\mathbb{R}^N)$, which, by Sobolev's Lemma, implies $(V_\ell \cdots V_1 \chi f_j)(0) \to (V_\ell \cdots V_1 f)(0)$. On the other hand, $\chi f_j \in \mathscr{S}(\mathbb{R}_z^N)$, hence $(V_\ell \cdots V_1 \chi f_j)(0) = 0$ which gives (A.5).

By induction we see that $V_\ell \cdots V_1(|z|^{-\gamma} T_\chi f)$ is a finite sum of terms of the form
$$g = h|z|^{-\gamma} V_{j_1} \cdots V_{j_k} T_\chi f$$
where $0 \leq k \leq \ell$ and $h \in IL_{R+1}^\infty(\mathbb{R}^N; \{0\})$. Since we have by (A.5)
$$V_{j_1} \cdots V_{j_k} T_\chi f = \chi V_{j_1} \cdots V_{j_k} f + g_0 = T_\chi V_{j_1} \cdots V_{j_k} f + g_0$$
with some smooth function $g_0 \in C_c^\infty(\mathbb{R}^N)$ vanishing near $z = 0$, our first step gives $g \in L^\infty(\mathbb{R}^N)$, and completes the proof. $\square$

We are now going to prove a converse of Proposition A.8. In fact, we need two steps.

LEMMA A.9. *Let $\gamma \in \mathbb{R}$ and $R > 0$ be arbitrary. If $s \in \mathbb{R}$ satisfies $s < \gamma + \frac{N}{2}$, then we have*
$$|z|^\gamma IL_R^\infty(\mathbb{R}^N; \{0\}) \subseteq |z|^s IL_R^2(\mathbb{R}^N; \{0\})$$
and the canonical map $|z|^\gamma IL_R^\infty(\mathbb{R}^N; \{0\}) \hookrightarrow |z|^s IL_R^2(\mathbb{R}^N; \{0\})$ is continuous.

Proof: Let $f = |z|^\gamma f_\gamma \in |z|^\gamma IL_R^\infty(\mathbb{R}^N; \{0\})$ be arbitrary. Then we have

$$\int_{\mathbb{R}_z^N} \left||z|^{-s} f(z)\right|^2 dz \leq \text{const} \int_0^R r^{2(\gamma-s)+N-1} dr = \text{const} \, \frac{R^{2(\gamma-s)+N}}{2(\gamma-s)+N} < \infty$$

because of $2(\gamma - s) + N > 0$.

As in the proof of Proposition A.8 we see by induction that for $V_j \in \mathcal{V}_E(\mathbb{R}^N; \{0\})$ the function $V_\ell \cdots V_1 |z|^{-s} f$ is a finite sum of terms of the form

$$h|z|^{-s}|z|^\gamma V_{j_1} \cdots V_{j_k} f_\gamma$$

with $0 \leq k \leq \ell$ and $h \in IL_{R+1}^\infty(\mathbb{R}^N; \{0\})$. Then the computation applies to $|z|^\gamma V_{j_1} \cdots V_{j_k} f_\gamma$ showing $V_\ell \cdots V_1 |z|^{-s} f \in L^2(\mathbb{R}^N)$; this completes the proof. $\square$

The second step is a little bit more complicated, and we start with a simple Lemma.

LEMMA A.10. *Let $R > 0$ and $f \in IL_R^2(\mathbb{R}^N; \{0\})$ be arbitrary. Then we have $\frac{z_j}{|z|} f \in IL_R^2(\mathbb{R}^N; \{0\})$ and $|z|\partial_{z_k} f \in IL_R^2(\mathbb{R}^N; \{0\})$ for $j, k = 1, \ldots, N$.*

Proof: The first statement follows by induction from $z_k \partial_{z_\ell} \frac{z_j}{|z|} = \frac{z_k}{|z|} \delta_{j,\ell} - \frac{z_j}{|z|} \frac{z_k}{|z|} \frac{z_\ell}{|z|}$, and the second one from the first and the identity $|z|\partial_{z_k} = \sum_{j=1}^N \frac{z_j}{|z|} z_j \partial_{z_k}$. $\square$

PROPOSITION A.11. *Let $R > 0$ and $s \geq 0$ be arbitrary. Then we have*

$$|z|^s IL_R^2(\mathbb{R}^N; \{0\}) \subseteq IH_R^s(\mathbb{R}^N; \{0\}).$$

Proof: Let $f \in IL_R^2(\mathbb{R}^N; \{0\})$ be arbitrary, and let $f_s := |z|^s f \in |z|^s IL_R^2(\mathbb{R}^N; \{0\})$ for $s \geq 0$. Let us first show $f_s \in IH_R^s(\mathbb{R}^N; \{0\})$ for all $s \in \mathbb{N}_0$. This is trivial for $s = 0$. Because of

$$\partial_{z_j} f_{s+1} = (s+1)|z|^s \left(\frac{z_j}{|z|} f\right) + |z|^s (|z|\partial_{z_j} f) \in H^s(\mathbb{R}^N)$$

for $j = 1, \ldots, N$ by induction and Lemma A.10, we have $f_{s+1} \in H^{s+1}(\mathbb{R}^N)$. Similarly to the proof of Proposition A.8 and Lemma A.9, we see that $V_\ell \cdots V_1 f_{s+1}$ is a finite sum of terms of the form

$$|z|^{s+1} V_{j_1} \cdots V_{j_k}(hf)$$

with $0 \leq k \leq \ell$ and $h \in IL_{R+1}^\infty(\mathbb{R}^N; \{0\})$. Because of $V_{j_1} \cdots V_{j_k}(hf) \in IL_R^2(\mathbb{R}^N; \{0\})$ we get $V_\ell \cdots V_1 f_{s+1} \in H^{s+1}(\mathbb{R}^N)$ which completes the inductional step.

Next, we are going to show $f_s \in IH_R^s(\mathbb{R}^N; \{0\})$ for all $0 < s < 1$ by complex interpolation. The general case follows then by combination.

We start with $f_s \in H^s(\mathbb{R}^N)$. Because of $H^s(\mathbb{R}^N) = [H^0(\mathbb{R}^N), H^1(\mathbb{R}^N)]_s$ by [**99**, Section 2.4.2] it suffices to show that there exists

$$F : \overline{S} := \{w \in \mathbb{C} : 0 \leq \operatorname{Re} w \leq 1\} \longrightarrow H^0(\mathbb{R}^N)$$

with the following properties [**99**, Section 1.9]

(a) $F : \overline{S} \longrightarrow H^0(\mathbb{R}^N)$ is continuous.
(b) $F|_S : S := \{w \in \mathbb{C} : 0 < \operatorname{Re} w < 1\} \longrightarrow H^0(\mathbb{R}^N)$ is analytic.
(c) $\sup_{w \in \overline{S}} e^{-|\operatorname{Im} w|} \|f(w)\|_{H^0(\mathbb{R}^N)} < \infty$.
(d) $\mathbb{R} \ni t \longmapsto F(1 + it) \in H^1(\mathbb{R}^N)$ is well-defined and continuous.
(e) $\max_{j=0,1} \sup_{t \in \mathbb{R}} e^{-|t|} \|F(j + it)\|_{H^j(\mathbb{R}^N)} < \infty$.
(f) $F(s) = f_s$.

Using Lebesgue's theorem on dominated convergence, Lemma A.10, and the cases $s = 0, 1$ we see that
$$F : \overline{S} \longrightarrow H^0(\mathbb{R}_z^N) : w \longmapsto [z \longmapsto |z|^w f(z)]$$
has all the desired properties.

To show $V_\ell \cdots V_1 f_s \in H^s(\mathbb{R}^N)$ we can argue as above. This completes the proof. □

We are now going to look what happens with the spaces $IL_R^\infty(\mathbb{R}^N; \{0\})$ after introducing polar coordinates $\beta_p : \overline{\mathbb{R}}_+ \times S^{N-1} \longrightarrow \mathbb{R}^N : (\varrho, \omega) \longmapsto z = \varrho\omega$ on $\mathbb{R}^N$. We need the following spaces:

Definition A.12. Let $R > 0$ and $\gamma \in \mathbb{R}$ be arbitrary. Then we define
$$\mathcal{C}_R^\infty(\overline{\mathbb{R}}_+ \times S^{N-1}; \partial) := \{f \in \mathcal{C}^{-\infty}(\overline{\mathbb{R}}_+ \times S^{N-1}) : \operatorname{sing\,supp} f \subseteq \{0\} \times S^{N-1},$$
$$f(r, \omega) = 0 \text{ for } r \geq R\},$$
$$IL_R^\infty(\overline{\mathbb{R}}_+ \times S^{N-1}; \partial) := \{f \in \mathcal{C}_R^\infty(\overline{\mathbb{R}}_+ \times S^{N-1}; \partial) :$$
$$\operatorname{Diff}_b^*(\overline{\mathbb{R}}_+ \times S^{N-1})f \subseteq L^\infty(\overline{\mathbb{R}}_+ \times S^{N-1})\},$$
$$\varrho^\gamma IL_R^\infty(\overline{\mathbb{R}}_+ \times S^{N-1}; \partial) := \{f \in \mathcal{C}^{-\infty}(\overline{\mathbb{R}}_+ \times S^{N-1}) : \varrho^{-\gamma} f \in IL_R^\infty(\overline{\mathbb{R}}_+ \times S^{N-1}; \partial)\}.$$

As for the spaces of Definition A.1 we see that there are natural Fréchet topologies on the spaces $IL_R^\infty(\overline{\mathbb{R}}_+ \times S^{N-1}; \partial)$ and $\varrho^\gamma IL_R^\infty(\overline{\mathbb{R}}_+ \times S^{N-1}; \partial)$.

Proposition A.13. *Let $R > 0$ and $\gamma \in \mathbb{R}$ be arbitrary. Then we have*
$$\beta_p^* \left(|z|^\gamma IL_R^\infty(\mathbb{R}^N; \{0\})\right) = \varrho^\gamma IL_R^\infty(\overline{\mathbb{R}}_+ \times S^{N-1}; \partial),$$
and the map $\beta_p^* : |z|^\gamma IL_R^\infty(\mathbb{R}^N; \{0\}) \longrightarrow \varrho^\gamma IL_R^\infty(\overline{\mathbb{R}}_+ \times S^{N-1}; \partial)$ is a topological isomorphism.

Proof: First note that the continuity of β_p^* and $(\beta_p^*)^{-1}$ follows from the closed graph theorem.

Let $f = \beta_p^* g \in \beta_p^*(|z|^\gamma IL_R^\infty(\mathbb{R}^N; \{0\}))$ be arbitrary. Then we have
$$W_\ell \cdots W_1(\varrho^{-\gamma} f) \in L^\infty(\overline{\mathbb{R}}_+ \times S^{N-1})$$
where $W_j \in \mathcal{V}_b(\overline{\mathbb{R}}_+ \times S^{N-1})$ is the lift of $V_j \in \mathcal{V}_E(\mathbb{R}^N; \{0\})$. Since these lifts generate $\mathcal{V}_b(\overline{\mathbb{R}}_+ \times S^{N-1})$ over $\mathcal{C}^\infty(\overline{\mathbb{R}}_+ \times S^{N-1})$ we have $f \in \varrho^\gamma IL_R^\infty(\overline{\mathbb{R}}_+ \times S^{N-1}; \partial)$.

On the other hand, if we have $f = \varrho^\gamma f_\gamma \in \varrho^\gamma IL_R^\infty(\overline{\mathbb{R}}_+ \times S^{N-1}; \partial)$ and define
$$g := (\beta_p)_* f : z \longmapsto f(|z|, \frac{z}{|z|}),$$
i.e. $\beta_p^* g = f$, then we get $[z \longmapsto |z|^{-\gamma} g(z)] = [z \longmapsto f_\gamma(|z|, \frac{z}{|z|})] \in L^\infty(\mathbb{R}^N)$. Since each $V_j \in \mathcal{V}_E(\mathbb{R}^N; \{0\})$ lifts to a $W_j \in \mathcal{V}_b(\overline{\mathbb{R}}_+ \times S^{N-1})$, we have
$$[z \longmapsto V_\ell \cdots V_1(|z|^{-\gamma} g)(z)] = [z \longmapsto (W_\ell \cdots W_1 f_\gamma)(|z|, \frac{z}{|z|})] \in L^\infty(\mathbb{R}^N)$$
which completes the proof. □

Bibliography

[1] L. Andersson, P. T. Chruściel, and H. Friedrich. On the regularity of solutions to the Yamabe equation and the existence of smooth hyperboloidal initial data for Einstein's field equations. *Comm. Math. Phys.*, 149:587–612, 1992.

[2] F. Baldus. $S(M,g)$-*pseudo-differential calculus with spectral invariance on* $\mathbb{R}^n$ *and manifolds for Banach function spaces*. PhD thesis, Universität Mainz, January 2001.

[3] R. Beals. Characterization of pseudodifferential operators and applications. *Duke Math. J.*, 44:45–57, 1977.

[4] D. Borthwick. Scattering theory and deformation of asymptotically hyperbolic metrics. Preprint, November 1997.

[5] D. Borthwick. Scattering theory for conformally compact metrics with variable curvature at infinity. *J. Funct. Anal.*, 184:313–376, 2001.

[6] J.-B. Bost. Principe d'Oka, K-théorie et systèmes dynamiques non commutatifs. *Invent. Math.*, 101:261–333, 1990.

[7] L. Boutet de Monvel. Boundary problems for pseudo-differential operators. *Acta Math.*, 126:11–51, 1971.

[8] B. H. Bowditch. Geometrical finiteness for hyperbolic groups. *J. Funct. Anal.*, 113:245–317, 1993.

[9] T. Bröcker and K. Jänich. *Einführung in die Differentialtopologie*. Springer-Verlag, Berlin, 1973. Heidelberger Taschenbücher, Band 143.

[10] R. Brooks, R. Gornet, and P. Perry. Isoscattering Schottky manifolds. *Geom. Funct. Anal.*, 10:307–326, 2000.

[11] R. R. Coifman and Y. Meyer. *Au delà des opérateurs pseudo-différentiels*, volume 57 of *Astérisque*. Société Mathématique de France, 1978.

[12] A. Connes. An analogue of the Thom isomorphism for crossed products of a C^*-algebra by an action of $\mathbb{R}$. *Adv. in Math.*, 39:31–55, 1981.

[13] H. O. Cordes. *On some C^*-algebras and Fréchet *-algebras of pseudo-differential operators*, volume 43 of *Proc. Symp. in Pure Math. – Pseudodifferential operators*, pages 79–104. Amer. Math. Soc., Providence, Rhode Island, 1985.

[14] J. Cuntz. Bivariante K-Theorie für lokalkonvexe Algebren und der Chern-Connes-Charakter. *Doc. Math.*, 2:139–182 (electronic), 1997.

[15] J. Dixmier. *Les C^*-algèbres et leur représentations*. Gauthier - Villars Éditeur, Paris, second french edition, 1969.

[16] J. Dunau. Fonctions d'un opérateur elliptique sur une variété compacte. *J. Math. Pures et Appl.*, 56:367–391, 1977.

[17] A. Dynin. Inversion problems for singular integral operators: C^*-approach. *Proc. Nat. Acad. Sci. U.S.A*, 75:4668–4670, 1978.

[18] Yu. V. Egorov and B.-W. Schulze. *Pseudodifferential operators, Singularities, Applications*, volume 93 of *Operator Theory and Applications*. Birkäuser, Basel, 1997.

[19] C. L. Epstein, R. B. Melrose, and G. A. Mendoza. Resolvent of the Laplacian on strictly pseudoconvex domains. *Acta Math.*, 167:1–106, 1991.

[20] I. C. Gohberg. On the theory of multidimensional singular integral operators. *Soviet Math.*, 1:960–963, 1960.

[21] C. R. Graham and J. M. Lee. Einstein metrics with prescribed conformal infinity on the ball. *Adv. Math.*, 87:186–225, 1991.

[22] B. Gramsch. Some homogeneous spaces in the operator theory and Ψ-algebras. In *Tagungsbericht Oberwolfach 42/81 – Funktionalanalysis: C^*-Algebren*, 1981.

[23] B. Gramsch. Relative Inversion in der Störungstheorie von Operatoren und Ψ-Algebren. *Math. Annalen*, 269:27–71, 1984.

[24] B. Gramsch. Fréchet algebras in the pseudodifferential analysis and an application to the propagation of singularities. In *Abstracts of the Conference* Partial Differential Equations Sept. 6–11, 1992, Preprint MPI/93-7, Bonn, 1992. MPI für Mathematik.

[25] B. Gramsch. Oka's principle for special Fréchet Lie groups and homogeneous manifolds in topological algebras of the microlocal analysis. In E. Albrecht and M. Mathieu, editors, *Banach algebras '97 – Proceedings of the 13th international conference on Banach algebras, Blaubeuren, July 20 – August 3, 1997*, pages 189–204, Berlin - New York, 1998. Walter de Gruyter.

[26] B. Gramsch. Fréchet algebras of pseudodifferential operators with asymptotic expansion are submultiplicative. in preparation.

[27] B. Gramsch and W. Kaballo. Multiplicative decompositions of holomorphic Fredholm functions and Ψ^*-algebras. *Math. Nachr.*, 204:83–100, 1999.

[28] B. Gramsch and K. G. Kalb. Pseudo-locality and hypoellipticity in operator algebras. Semesterbericht Funktionalanalysis, Universität Tübingen, Sommersemester 1985. 51–61.

[29] B. Gramsch and E. Schrohe. *Submultiplicativity of Boutet de Monvel's algebra for boundary value problems*, pages 235–258. Mathematical topics – Advances in Partial Differential Equations. Akademie-Verlag, Berlin, 1994.

[30] B. Gramsch, J. Ueberberg, and K. Wagner. Spectral invariance and submultiplicativity for Fréchet algebras with applications to pseudo-differential operators and Ψ^*-quantization. In *Operator Theory: Advances and Applications*, vol. 57, pages 71–98. Birkhäuser, Basel, 1992.

[31] B. Gérard. *Singular connections on three-manifolds and manifolds with cylindrical ends*. PhD thesis, Brandeis University, 1997. math.DG/9805036.

[32] D. Grieser. Basics of the b-calculus. In *Approaches to singular analysis (Berlin, 1999)*, pages 30–84. Birkhäuser, Basel, 2001.

[33] L. Guillopé and M. Zworski. Scattering asymptotics for Riemann surfaces. *Ann. of Math.*, 145:597–660, 1997.

[34] P. D. Hislop. The geometry and spectra of hyperbolic manifolds. *Proc. Indian Acad. Sci. Math. Sci.*, 104:715–776, 1994. Spectral and inverse spectral theory (Bangalore, 1993).

[35] L. Hörmander. Fourier integral operators I. *Acta Math.*, 127:79–183, 1971.

[36] L. Hörmander. *The analysis of linear partial differential operators*, vol. 3. *Pseudodifferential operators*, volume 274 of *Grundlehren der Mathematischen Wissenschaften*. Springer-Verlag, Berlin - Heidelberg - New York, 1985.

[37] M. S. Joshi and A. Sá Barreto. Inverse scattering on asymptotically hyperbolic manifolds. *Acta Math.*, 184:41–86, 2000.

[38] R. Lauter. *Holomorphic functional calculus in several variables and Ψ^*-algebras of totally characteristic operators on manifolds with boundary*. PhD thesis, Johannes Gutenberg Universität-Mainz, November 1996.

[39] R. Lauter. O predstavleniyah Ψ^*- i C^*- algebr psevdodifferentsial'nyh operatorov na mnogoobraziyah s vershinami. In N. N. Ural'tseva, editor, *Nelineinye uravneniya s chastnymi proizvodnymi i teoriya funktsii*, volume 18 of *Problemy matematicheskogo analiza*, pages 85–117. Sankt-Peterburgskii universitet, November 1998. in Russian.

[40] R. Lauter. An operator theoretical approach to enveloping Ψ^*- and C^*-algebras of Melrose algebras of totally characteristic pseudodifferential operators. *Math. Nachr.*, 196:141–166, 1998.

[41] R. Lauter. On the existence and structure of Ψ^*-algebras of totally characteristic operators on compact manifolds with boundary. *J. Funct. Anal.*, 169:81–120, 1999.

[42] R. Lauter. The length of C^*-algebras of b-pseudodifferential operators. *Proc. of the Amer. Math. Soc.*, 128:1955–1961, 2000.

[43] R. Lauter, B. Monthubert, and V. Nistor. Pseudodifferential analysis on continuous family groupoids. *Doc. Math.*, 5:625–655 (electronic), 2000.

[44] R. Lauter and S. Moroianu. Fredholm theory for degenerate pseudodifferential operators on manifolds with fibered boundaries. *Comm. Partial Differential Equations*, 26:233–283, 2001.

[45] R. Lauter and S. Moroianu. The index of cusp operators on manifolds with corners. *Ann. Global Anal. Geom.*, 21:31–49, 2002.

[46] R. Lauter and V. Nistor. Analysis of geometric operators on open manifolds: a groupoid approach. In N.P. Landsman, M. Pflaum, and M. Schlichenmaier, editors, *Quantization of Singular Symplectic Quotients*, volume 198 of *Progress in Mathematics*, pages 181–229. Birkhäuser, Basel - Boston - Berlin, 2001.

[47] P. D. Lax and R. S. Phillips. *Scattering theory for automorphic functions*. Princeton Univ. Press, Princeton, N.J., 1976. Annals of Mathematics Studies, No. 87.

[48] C. R. LeBrun. $\mathcal{H}$-space with a cosmological constant. *Proc. Roy. Soc. London Ser. A*, 380:171–185, 1982.

[49] J. M Lee. Fredholm operators and Einstein metrics on conformally compact manifolds. Preprint, May 2001.

[50] P. A. Loya. *On the b-pseudodifferential calculus on manifolds with corners*. PhD thesis, Massachusetts Institute of Technology, Cambridge, MA, 1998.

[51] P. A. Loya. *b*-pseudodifferential operators on manifolds with corners. book: in preparation, 2001.

[52] F. Mantlik. Norm closure and extension of the symbolic calculus for the cone algebra. *Ann. Global Anal. and Geometry*, 13:339–376, 1995.

[53] R. R. Mazzeo. *Hodge Cohomology of negatively curved manifolds*. PhD thesis, Massachusetts Institute of Technology, Cambridge, MA, 1986.

[54] R. R. Mazzeo. The Hodge cohomology of a conformally compact metric. *J. Differential Geometry*, 28:309–339, 1988.

[55] R. R. Mazzeo. Elliptic theory of differential edge operators I. *Comm. Partial Differential Equations*, 16:1615–1664, 1991.

[56] R. R. Mazzeo. Unique continuation at infinity and embedded eigenvalues for asymptotically hyperbolic manifolds. *Amer. J. Math.*, 113:25–45, 1991.

[57] R. R. Mazzeo. Edge operators in geometry. In *Symposium "Analysis on Manifolds with Singularities" (Breitenbrunn, 1990)*, pages 127–137. Teubner, Stuttgart, 1992.

[58] R. R. Mazzeo and R. B. Melrose. Meromorphic extension of the resolvent on complete spaces with asymptotically constant negative curvature. *J. Funct. Anal.*, 75:260–310, 1987.

[59] R. R. Mazzeo and R. B. Melrose. Analytic surgery and the eta invariant. *Geometric and Functional Analysis*, 5:14–75, 1995.

[60] R. R. Mazzeo and R. B. Melrose. Pseudodifferential operators on manifolds with fibred boundaries. *Asian J. Math.*, 2:833–866, 1998.

[61] R. R. Mazzeo and R. S. Phillips. Hodge theory on hyperbolic manifolds. *Duke Math. J.*, 60:509–559, 1990.

[62] R. B. Melrose. Analysis on manifolds with corners. book: in preparation.

[63] R. B. Melrose. Introduction to microlocal analysis. Lecture Notes – Spring 1990, Massachusetts Institute of Technology.

[64] R. B. Melrose. Transformation of boundary value problems. *Acta Math.*, 147:149–236, 1981.

[65] R. B. Melrose. Analysis on manifolds with corners. Lectures at Massachusetts Institute of Technology from notes taken by B. T. Livingston and M. R. Zworski, July 1988.

[66] R. B. Melrose. Pseudodifferential operators, corners and singular limits. In *Proceeding of the International Congress of Mathematicians, Kyoto*, pages 217–234, Berlin - Heidelberg - New York, 1990. Springer-Verlag.

[67] R. B. Melrose. Calculus of conormal distributions on manifolds with corners. *Internat. Math. Res. Notices*, 3:51–61, 1992.

[68] R. B. Melrose. *The Atiyah-Patodi-Singer index theorem*, volume 4 of *Research Notes in Mathematics*. A K Peters, Wellesley, Massachusetts, 1993.

[69] R. B. Melrose. Spectral and scattering theory for the Laplacian on asymptotically Euclidean space. In M. Ikawa, editor, *Spectral and Scattering Theory*, volume 162 of *Lecture Notes in Pure and Applied Mathematics*, pages 85–130, New York, 1994. Marcel Dekker Inc. Proceedings of the Taniguchi International Workshop held in Sanda, November 1992.

[70] R. B. Melrose. *Geometric scattering theory*. Cambridge University Press, 1995.

[71] R. B. Melrose. Geometric optics and the bottom of the spectrum. In F. Colombini and N. Lerner, editors, *Geometrical optics and related topics*, volume 32 of *Progress in nonlinear differential equations and their applications*. Birkhäuser, Basel - Boston - Berlin, 1997.

[72] R. B. Melrose and V. Nistor. Homology of pseudodifferential operators I. Manifolds with boundary. to appear in: Amer. Math. J.; Preprint, May 1996.

[73] R. B. Melrose and P. Piazza. Analytic K-theory on manifolds with corners. *Adv. Math.*, 92:1–26, 1992.

[74] R. B. Melrose and J. Wunsch. Singularities and the wave equation on conic spaces. In *Geometric analysis and applications (Canberra, 2000)*, pages 170–182. Austral. Nat. Univ., Canberra, 2001.

[75] W. Müller. Spectral geometry and scattering theory for certain complete surfaces of finite volume. *Invent. Math.*, 109:265–305, 1992.

[76] B. Monthubert and F. Pierrot. Indice analytiques et groupoïdes de Lie. *C. R. Acad. Sci. Paris Ser. I*, 325(2):193–198, 1997.

[77] M. A. Neumark. *Normierte Algebren*. VEB Deutscher Verlag der Wissenschaften, second german edition, 1990.

[78] V. Nistor. Groupoids and the integration of Lie algebroids. *J. Math. Soc. Japan*, 52:847–868, 2000.

[79] V. Nistor, A. Weinstein, and P. Xu. Pseudodifferential operators on groupoids. *Pacific J. Math.*, 189:117–152, 1999.

[80] N. C. Phillips. K-theory for Fréchet algebras. *Internat. J. of Math.*, 2:77–129, 1991.

[81] B. A. Plamenevskij. On algebras generated by pseudodifferential operators with isolated singularities in the symbols. *Sel. Math. Sov.*, 5:77–100, 1986. Originally published in: *Problemy Matematicheskoi fiziki*, No. 10 Spektralnaya Teoriya, volnovoye protsessy, Leningrad University Press, 1982, pp. 209–241.

[82] B. A. Plamenevskij. *Algebras of Pseudodifferential Operators*. Kluwer Academic Publishers, Dordrecht - Boston - London, 1989. Originally published in Russian, *Nauka*, Moscow, 1986.

[83] B. A. Plamenevskij and V. N. Senichkin. Solvable algebras of operators. *St. Petersburg Math. J.*, 6:895–968, 1995. Originally published in Russian – *Algebra i Anal.*, 1994.

[84] B. A. Plamenevskij and V. N. Senichkin. Pseudodifferential operators on manifolds with singularities. *Funktsional. Anal. i Prilozhen.*, 33:88–91, 1999.

[85] B. A. Plamenevskij and V. N. Senichkin. On a class of pseudodifferential operators on R^m and on stratified manifolds. *Mat. Sb.*, 191:109–142, 2000.

[86] J. Råde. Elliptic uniformly degenerate operators. Preprint, February 1999.

[87] J. Råde. Singular Yang-Mills fields. Local theory. I. *J. Reine Angew. Math.*, 452:111–151, 1994.

[88] E. Schrohe. A Ψ^*-algebra of pseudo-differential operators on noncompact manifolds. *Arch. Math.*, 51:81–86, 1988.

[89] E. Schrohe. Fréchet algebra techniques for boundary value problems: Fredholm criteria and functional calculus via spectral invariance. *Math. Nachr.*, 199:145–185, 1999.

[90] B.-W. Schulze. Pseudo-differential operators on manifolds with edges. In B.-W. Schulze and H. Triebel, editors, *Proceedings of the International Symposium held in Holzhau, April 25-29, 1988*, pages 259–287, Leipzig, 1989. BSB B. G. Teubner Verlagsgesellschaft.

[91] B.-W. Schulze. *Pseudo-differential operators on manifolds with singularities*. North-Holland, Amsterdam - London - New York - Tokyo, 1991.

[92] B.-W. Schulze. *Pseudo-differential boundary value problems, conical singularities, and asymptotics*. Akademie Verlag, Berlin, 1994.

[93] B.-W. Schulze. Boundary value problems and edge pseudo-differential operators. In *Microlocal analysis and spectral theory (Lucca, 1996)*, pages 165–226. Kluwer Acad. Publ., Dordrecht, 1997.

[94] B.-W. Schulze and J. Seiler. The edge algebra of boundary value problems. Preprint, April 2001.

[95] V. N. Senichkin. Pseudodifferential operators on manifolds with edges. *J. Math. Sci.*, 73:711–747, 1995. Differential and pseudodifferential operators.

[96] V. N. Senichkin. The spectrum of the algebra of pseudodifferential operators on a manifold with smooth edges. *Algebra i Analiz*, 8:105–147, 1996.

[97] S. R. Simanca. *Pseudo-differential operators*, volume 171 of *Pitman research notes in mathematics*. Longman Scientific & Technical, Harlow, Essex, 1990.

[98] M. Traute. Spezielle Fréchet Algebren für singuläre Integraloperatoren mit Unstetigkeiten in den Symbolen. Diplomarbeit, Fachbereich 17-Mathematik, Johannes Gutenberg-Universität, Mainz, 1989.

[99] H. Triebel. *Interpolation theory, function spaces, differential operators*. North-Holland Publishing Company, Amsterdam - New York - Oxford, 1978.

[100] J. Ueberberg. Zur Spektralinvarianz von Algebren von Pseudodifferentialoperatoren in der L^p-Theorie. *Manuscripta Math.*, 61:459–475, 1988.

Notations

b-geometry

    B_1 4
    $B_{1,O}$ 7
    $B_{1,T}$ 7
    Δ_b 4, 19
    M_b^2 19
    X_b^3 8
    X_b^2 4
    β_b^3 8
    ^bTZ xv
    β_b^2 .... 4, 19
    ff^b 5
    $\mathrm{ff}^b(j)$ 19
    j^b xv
    lb 4
    rb 4
    $\mathcal{V}_b(Z)$ xv

b-c-geometry

    M 19
    Q_t 25
    $A^\sharp$ 23
    $\Delta_{b,c}$ 20
    $F_{M_N,-}$ 20
    $F_{M_N,+}$ 20
    M_N 20
    M_b^2 19
    $M_{b,c}^2$ 19
    $\widetilde{M}_{b,c}^2$ 20
    $\widetilde{M}_b^2$ 20
    $KD_{b,c}^{1/2}$ 21
    $^{b,c}\Omega^\alpha(M)$ .. 21
    $^{b,c}TM$ 19
    $^{b,c}T^*M$ 19
    $^{b,c}\overline{T}^*M$ 22
    β_b^2 19
    ff^{M_N} 20
    $\mathrm{ff}^b(0)$ 20
    $\mathrm{ff}^b(j)$ 19
    ff^c 20
    $\gamma' := (\gamma'_{\mathrm{lb}}, \gamma'_{\mathrm{rb}})$ 22
    $\gamma = (\gamma_{\mathrm{lb}}, \gamma_{\mathrm{rb}})$ 22
    $j^{b,c}$ 19
    lb 20
    $\mathrm{lb}(j)$ 22
    $\mathrm{lb} = \mathrm{lb}(0)$ 20

rb 20
$\mathrm{rb}(j)$ 22
$\mathrm{rb} = \mathrm{rb}(0)$ 20
ϱ_1 19
ϱ_0 19
$\varphi_{b,c}$ 20
φ_b 20
$\mathcal{V}_{b,c}(M)$ 19

0-geometry

    $<\cdot,\cdot>_{L^2(X,{}^0\Omega^{1/2})}$ 10
    B_0 7
    $N\Delta_{0,e}$ 12
    $V^{\partial X}$ 29
    $A^\sharp$ 13
    B_O 8
    B_1 4
    $B_{1,O}$ 7
    $B_{1,T}$ 7, 8
    B_0 7
    $B_{0,e}$ 5
    B_2 5
    $B_{2,O}$ 8
    $B'_{2,O}$ 8
    $\Delta_{0,e}$ 5
    $F_{0,e}$ 5
    T_0 7
    $T_{0,e}$ 5
    $X_{0,e}^3$ 8
    X_0^2 .... xv, 4, 6
    $X_{0,e}^2$ .... xv, 4, 5
    $\beta_{0,e}^3$ 8
    β_0^2 .... 4, 6
    $\beta_{0,e}^2$ 5
    ff^0 7
    $\mathrm{ff}^{0,e}$ 5
    $\gamma = (\gamma_T, \gamma_B, \gamma_F)$ 16
    $KD_{0,e}^{1/2}$ 10
    $^0\Omega^\alpha(X)$ 9
    $^0S^*X$ 13
    ^0TX 3
    $^0T\partial X$ 4
    $^0T^*X$ 3
    $^0\overline{T}^*X$ 13
    $\pi_{L;0,e}$ 11
    $\pi_{R;0,e}$ 11

π_C 7
π_F 7
π_O 7
π_S 7
$\pi_O^{0,e}$ 8
$\mathcal{V}_0(X)$ .... ix, 3
j_{ϱ_N} 34

b-c-calculus

$(\cdot)_{\partial_j}$ 23
Q 26
$I_b(A)$ 23
$I_c^{(k)}(A)$ 23
$\mathcal{L}(\varrho_0^{\mathfrak{a}_0}\varrho_1^{\mathfrak{a}_1}L^2(M,{}^{b,c}\Omega^{1/2}))$ 26
$L^2(M,{}^{b,c}\Omega^{1/2})$ 21
$\varrho_0^{\mathfrak{a}_0}\varrho_1^{\mathfrak{a}_1}L^2(M,{}^{b,c}\Omega^{1/2})$ 26
$\Psi_{b,c}^m(M;{}^{b,c}\Omega^{1/2})$ 21
$\Psi_{b,c}^{m,k,\gamma}(M;{}^{b,c}\Omega^{1/2})$ 22
$\widetilde{\Psi}_{b,c}^{-\infty,-\infty,\gamma}(M,{}^{b,c}\Omega^{1/2})$ 22
$\Psi^{-\infty,\gamma}(M,{}^{b,c}\Omega^{1/2})$ 22
$\Psi_{b,c}^{m,k,\gamma}(M;E,F)$ 28
${}^{b,c}\sigma^{(m,k)}$ 23
$\mathcal{K}(\varrho_0^{\mathfrak{a}_0}\varrho_1^{\mathfrak{a}_1}L^2(M,{}^{b,c}\Omega^{1/2}))$ 26
$\mathcal{K}^{-\infty,\gamma}(M_{b,c}^2;KD_{b,c}^{1/2})$ 22
$\mathcal{R}_{\varrho_j}$ 23
$\tau_{b,c}$ 26

0-calculus

$\Psi_0^{m,k,\gamma}(X;E,F)$ 18
$(\cdot)_\partial$ 17
$<\cdot,\cdot>_{L^2(X,{}^0\Omega^{1/2})}$ 45
$H_0^0(X,E)$ 48
$I^m(X_{0,e}^2,\Delta_{0,e};KD_{0,e}^{1/2})$ 12
I_A 18
$I_{cl}^m(X_{0,e}^2,\Delta_{0,e};KD_{0,e}^{1/2})$ 11
$\mathcal{F}_{\varrho_N}^\nu$ 31
$\mathcal{F}_{\varrho_N}^{\nu,\chi}$ 30
$\varrho_N^{\mathfrak{a}}H_0^s(X,{}^0\Omega^{1/2})$ 47
I_{M_N} 34
$\mathcal{L}(\varrho_N^{\mathfrak{a}}L^2(X,{}^0\Omega^{1/2}))$ 47
$\Lambda_{m,k}$ 53
$\Lambda_{m,k}^{s,\mathfrak{a}}$ 53
$L^2(X,{}^0\Omega^{1/2})$ 45
$L^2(X,E)$ 48
$\varrho_N^{\mathfrak{a}}L^2(X,{}^0\Omega^{1/2})$ 47
$\mathcal{N}_{\varrho_N}^\nu$ 31
$\mathcal{N}_{\varrho_N}^{\nu,\chi}$ 30
$\mathfrak{N}_0^{-\infty,0,\gamma}$ 38

$\widetilde{\mathfrak{N}}_0^{-\infty,0,(\gamma_T,\gamma_B)}(\delta)$ 37
$\Psi_0^m(X;{}^0\Omega^{1/2})$ 11
$\Psi_0^{m,k}(X;{}^0\Omega^{1/2})$ 14
$\Psi_0^{m,k,\gamma}(X;{}^0\Omega^{1/2})$ 16
$\Psi_0^{-\infty,k,\gamma}(X;{}^0\Omega^{1/2})$ 16
$\mathcal{P}^*\partial X$ 36
$\mathcal{E}$ 31
$\mathcal{E}_0$ 57
$\mathcal{N}^{(k)}$ 42
$\mathcal{R}_{\varrho_N}$ 17
$\mathfrak{N}$ 32
γ_z 17
$\mathcal{K}_0^{-\infty,k,\gamma}(X_{0,e}^2,KD_{0,e}^{1/2})$ ... 16
${}^0\sigma^{(m)}$ 12
${}^0\sigma^{(m,k)}$ 15
ϱ_0 37
ϱ_∞ 37
τ_0 58

C^*- and Ψ^*-algebras

$I_b^{\mathfrak{a}_0}$ 27
I_c 27
J_1 60
J_2 60
$Q^{(\mathfrak{a})}$ 58
$\mathcal{A}_{b,c}^{(\mathfrak{a}_0,\mathfrak{a}_1)}(M,{}^{b,c}\Omega^{1/2})$ 71
$\mathcal{A}_0^{(\mathfrak{a})}(X,{}^0\Omega^{1/2})$ 71
$\mathcal{B}_{b,c}^{(\mathfrak{a})}(M,{}^{b,c}\Omega^{1/2};S^*\partial X)$ 59
$\mathcal{B}_{b,c}^{(\mathfrak{a}_0,\mathfrak{a}_1)}(M,{}^{b,c}\Omega^{1/2})$ 27
$\mathcal{B}_{b,c}^{(\mathfrak{a})}(M,{}^{b,c}\Omega^{1/2})$ 58
$\mathcal{B}_0^{(\mathfrak{a})}(X,{}^0\Omega^{1/2})$ .... xiv, 58
$\mathcal{L}(\varrho_0^{\mathfrak{a}_0}\varrho_1^{\mathfrak{a}_1}L^2(M,{}^{b,c}\Omega^{1/2}))$ 26
$\mathcal{L}(\varrho_N^{\mathfrak{a}}L^2(X,{}^0\Omega^{1/2}))$ 47
$\mathcal{N}_{\varrho_N}^{\nu,\mathfrak{a}}$ 58
Φ 64
Ψ 67
Ψ^V 70
Ψ_n^V 70
$Q_{b,c}^{(\mathfrak{a}_0,\mathfrak{a}_1)}$ 27
$Q_0^{(\mathfrak{a})}$ 58, 62
$\mathcal{T}_\mathcal{J}(\mathcal{B})$ 57
${}^{b,c}\sigma^{(0,0)}$ 27
$\mathcal{A}^{-1}$ 69
$\mathcal{B}(;\Gamma_{\mathfrak{a}_0})$ 26
$\mathcal{B}(;\mathbb{R})$ 26
$\mathcal{B}^{-1}$ 69

$\mathcal{I}_2$ 63
$\mathcal{I}_3$ 63
$\mathcal{J}_0$ 27
$\mathcal{J}_1$ 63
$\mathcal{J}_2$ 63
$\mathcal{K}(\varrho_0^{\mathfrak{a}_0} \varrho_1^{\mathfrak{a}_1} L^2(M,^{b,c}\Omega^{1/2}))$ 26
$\mathcal{M}$ 68
$\mathcal{C}_b(\Gamma_{\mathfrak{a}_0})$ 26
$\mathcal{C}_b(\mathbb{R})$ 26
δ_V 70
$\mathfrak{M}$ 64
$\mathfrak{U}(\mathfrak{m})$ 64
$\mathfrak{V}_q$ 67
$^0\sigma^{(0)}$ 58
$\sigma_\mathcal{B}^{(0)}$ 26
τ_0 58
$\widehat{\mathcal{B}_0^{(\mathfrak{a})}(X,{}^0\Omega^{1/2})}$ 63
$\widehat{\mathcal{Q}_0^{(\mathfrak{a})}}$ 63
$\widehat{\mathcal{B}}$ 57
$\widetilde{\mathcal{A}}$ 72

conormal functions
T_χ 76
$|z|^\gamma IL_R^\infty(\mathbb{R}^N;\{0\})$ 75
$|z|^\gamma IL_R^2(\mathbb{R}^N;\{0\})$ 75
$\mathcal{C}_R^\infty(\mathbb{R}^N;\{0\})$ 73
$\mathcal{C}_R^\infty(\overline{\mathbb{R}}_+ \times S^{N-1};\partial)$ 78
$IH_R^s(\mathbb{R}^N;\{0\})$ 73
$IH_c^s(\mathbb{R}^N;\{0\})$ 73
$IL_R^\infty(\mathbb{R}^N;\{0\})$ 73, 75
$IL_R^\infty(\overline{\mathbb{R}}_+ \times S^{N-1};\partial)$ 78
$IL_c^\infty(\mathbb{R}^N;\{0\})$ 73
$IL_R^2(\mathbb{R}^N;\{0\})$ 73, 75
$IL_c^2(\mathbb{R}^N;\{0\})$ 73
$I_R^m(\mathbb{R}^N;\{0\})$ 74
$I_c^m(\mathbb{R}^N;\{0\})$ 74
$\varrho^\gamma IL_R^\infty(\overline{\mathbb{R}}_+ \times S^{N-1};\partial)$ 78
$r^\gamma L^2(\overline{\mathbb{R}}_+;dr)$ 75

vector fields
$\mathcal{V}_\mathcal{H}(Z)$ 16
$\mathcal{M}_{\mathrm{lin}}(\mathbb{R}^N)$ 73
$\mathcal{V}(X)$ 3
$\mathcal{V}_b(Z)$ xv
$\mathcal{V}_b(X)$ x
$\mathcal{V}_{b,c}(M)$ 19
$\mathcal{V}_c(X)$ x
$\mathcal{V}_e(X)$ x
$\mathcal{V}_E(\mathbb{R}^N;\{0\})$ 73

$\mathcal{V}_0(X)$ .... ix, 3

further notations
F_s 58
$I_{cl}^m(X,Y;E)$ 11
$N(T)$ xvi
$R(T)$ xvi
$S^{[m]}(\mathbb{R})$ 24
$S^m(;\mathbb{R}_\zeta^N)$ 73
$T^+X_b^2$ 5
X 3
$[\mathfrak{g},\mathfrak{g}]$ 3
$A^\sharp$ 13
$|\cdot|_s$ 58
$\Gamma_{\mathfrak{a}_0}$ 26
Γ 29
$\mathcal{M}_k(Z)$ xv
$M_\mathcal{O}^m$ 23
$N^+\partial X$ 29
$\Omega(Z)$ xv
Φ_{ϱ_N} 29
RC 20
$\mathscr{S}(E)$ xvi
$S^{[m]}(E)$ xv
$S_{cl}^m(;\mathbb{R}_\xi)$ 23
$S_{cl}^0(;\mathbb{R})$ 69
$S^*\partial X$ 37
$\overline{T}^*\partial X$ 36
$\mathfrak{a}_1$ 26
$\mathfrak{a}_0$ 26
∂X 3
β_p 39, 75, 78
$\mathcal{A}_\mathcal{H}^\gamma(Z)$ 16
$\mathcal{B}(V)$ 70
$\mathcal{D}(V)$ 70
$\mathcal{I}_q$ 3
$\mathcal{C}_b(\mathbb{R})$ 69
$\mathcal{C}^{-\infty}(Z,E)$ xv
$\dot{\mathcal{C}}^\infty(Z,E)$ xv
$\mathfrak{g}$ 3
$\mathfrak{g}^{(k)}$ 3
$\int_X$ 9
κ_A 11
$\mathcal{M}_{\mathrm{lin}}(\mathbb{R}^N)$ 73
ν 29
Ω^α 9
Ω_{fiber} 12
π xvi
ϱ_N 3

ϱ_H xv
$\sigma^{(m)}$ 24
$\sigma^{(0)}$ 26
S_+^{n-1} 5
$\widehat{\cdot}$ 58
$s: S^*\partial X \to T^*\partial X \setminus \{0\}$ 58

Index

0-α-densities........................9
0-calculus with bounds...........16
0-cosphere bundle................13
0-cotangent bundle................3
0-diagonal........................5
0-differential operator.........ix, 42
0-double space..............xv, 4, 6
0-front face......................7
0-metric.........................ix
0-pseudodifferential operator......11
0-structure.......................x
0-tangent bundle..................3
0-vector fields..............ix, xi, 3
Ψ^*-algebra...................xiii, 69
Ψ^*-completions...................71
Θ-pseudodifferential operator.....13
α-densities........................9
$\mathbb{R}_+$-action....................25
b-blow up........................4
b-calculus......................xv
b-diagonal...................4, 19
b-front face..................5, 20
b-indicial family.................23
b-structure.......................x
b-tangent bundle.................xv
b-triple space....................8
c-indicial family.................23
b-c-α-densities....................21
b-c-blow up.....................19
b-c-cosphere bundle...............22
b-c-cotangent bundle..............19
b-c-diagonal.....................20
b-c-double space..................19
b-c-kernel half densities...........21
b-c-operator.....................xii
b-c-tangent bundle................19
b-c-vector fields..................19

algebra
   Ψ^*-algebra..............xiii, 69
   essentially commutative......27
   length of a C^*-algebra.......57
   solvable C^*-algebra......xiv, 57
   spectrum of a C^*-algebra....57
   spectrum of an algebra......xiv
asymptotic completeness..........15

Atiyah-Bott obstruction..........xiii

blow up
   0-blow up......................6
   b-blow up......................4
   b-c-blow up...................19
   extended 0-blow-up...........5
   of zero section...............36
blowing up........................4
bottom face.......................5
boundary
   left boundary.............4, 20
   right boundary............4, 20
boundary faces...................xv
boundary fibration structure.......x
   0-structure....................x
   b-structure....................x
   cusp-structure.................x
   edge-structure.............x, xi
boundary hyperface..............xv
boundary value problem......xi, xiii
Boutet de Monvel calculus....xi, xiii

calculus
   0-calculus with bounds.......16
   b-calculus....................xv
   b-c-calculus with bounds.....22
   Boutet de Monvel calculus..xiii
   extended 0-calculus..........14
   extended b-c-calculus.........22
   full calculus...................x
   small 0-calculus..............11
   small b-c-calculus............21
   small calculus..................x
characterization
   Fredholm property
      $\Psi_0^{m,k}(X;{}^0\Omega^{1/2})$............54
      $\Psi_0^0(X;{}^0\Omega^{1/2})$..............50
      b-c-calculus................24
   reduced normal operator
      $\Psi_0^m(X;{}^0\Omega^{1/2})$.............34
      $\Psi_0^{-\infty}(X;{}^0\Omega^{1/2})$...........32
      $\Psi_0^{-\infty,0,\gamma}(X;{}^0\Omega^{1/2})$.........38
codimension of a boundary face...xv
commutator ideal of a Lie algebra..3
compactification.................xvi

radial 13, 20, 22, 36
completeness
    asymptotic 15
composition series 57
    solving of length ℓ 57
conformal infinity ix
conformally compact space ix
conical set 64
conormal distributions 11
conormal function 75
coordinates
    polar coordinates 6
    projective coordinates 6
cosphere bundle
    0-cosphere bundle 13
    b-c-cosphere bundle 22
    of the boundary 37
cotangent bundle
    0-cotangent bundle 3
    b-c-cotangent bundle 19
cusp front face 20
cusp-structure x

defining function xv
densities xv
    0-α-densities 9
    0-kernel half-densities 10
    α-densities 9
    b-c-α-densities 21
    b-c-kernel half densities 21
    fiber density 12
derivation 70
diagonal
    b-diagonal 4, 19
    b-c-diagonal 20
    extended 0-diagonal 5
double space
    0-double space .... xv, 4, 6
    b-c-double space 19
    extended 0-double space xv, 4, 5

edge-structure .... x, xi
elliptic operator 15, 42
elliptic regularity 71
essentially commutative 27
extended 0 blow-up 5
extended 0-calculus 14
extended 0-double space ..... xv, 4, 5
extended 0-front face 5

extended 0-kernel half-densities ... 10
extended 0-triple space 8
extended b-c-calculus 22
extended b-c-calculus with bounds 22
extendible distributions .... xv, 11

face
    bottom face 5
    front face
        0-front face 7
        b-front face 5, 20
        cusp front face 20
        extended 0-front face 5
    top face 5
fiber density 12
formal adjoint 13, 23
Fréchet-topology 12
Fredholm theory
    for 0-operators 50, 54
    for b-c-operators 24
full calculus x

groupoid .... xi, xiv

Hilbert spaces
    scale of 71
homogeneous principal part 24, 26, 33
homogeneous principal symbol 12
hyperbolic space ix
hyperface xv

indicial family
    b-indicial family 23
    c-indicial family 23
indicial function .... xii, 18
inverse-closed 25
irreducible representation 57

Jacobson topology .... xiv, 57
joint 0-symbols 58, 62
joint b-c-symbols 27

length of a C^*-algebra 57
Lie algebroid xi
lifted Schwartz kernel 11, 21
linear vector fields 73
Lopatinskij-Shapiro xiii

manifold with boundary 3
manifold with corners xv

INDEX

maximal ideal space 68

normal fibration 29
normal operator xi

operator
    0-differential operator ..... ix, 42
    b-c-operator xii
    Bessel type operator xii
    elliptic 15, 42
    Laplace-Beltrami ix
    normal operator xi
    order reducing 54
    reduced normal operator xii, xiii
    uniformly degenerated ix
order-reducing operators 54
oscillatory testing 49

polar coordinates 6, 75, 78
positive normal bundle 29
principal symbol xi
    for 0-calculus 12
    for b-c-calculus 23
    for conormal distributions 12
    for extended 0-calculus 15
projective coordinates 6
pseudodifferential operators
    0-calculus xii
    0-calculus with bounds 16
    b-c-calculus with bounds 22
    small 0-calculus 11
    small b-c-calculus 21
push-forward 13

radial compactification ... xvi, 13, 20, 22, 36
reduced normal operator . xii, xiii, 30, 31, 42
    characterization
        $\Psi_0^m(X; {}^0\Omega^{1/2})$ 34
        $\Psi_0^{-\infty}(X; {}^0\Omega^{1/2})$ 32
        $\Psi_0^{-\infty,0,\gamma}(X; {}^0\Omega^{1/2})$ 38
representation
    irreducible 57
restriction map 17
restriction to the boundary ... 17, 23

scale of Hilbert spaces 71
Schwartz kernel 4

Schwartz kernel theorem 11, 21
Schwartz space xvi
seminorm
    submultiplicative 69
small 0-calculus 11
small b-c-calculus 21
small calculus x
Sobolev spaces
    weighted 0-Sobolev spaces 47
    weighted b-c-Sobolev spaces .. 24
solvable C^*-algebra .... xiv, 57
solvable Lie algebra 4
solving composition series 57
space
    conformally compact ix
    hyperbolic ix
spectral invariance 69
spectrum xiv
spectrum of a C^*-algebra 57
submultiplicative 69
submultiplicative seminorms xiii
symbol 73
symbol morphism 26, 53
symbol reproducing families 49
symbolic structure 26, 53

tangent bundle
    0-tangent bundle 3
    b-tangent bundle xv
    b-c-tangent bundle 19
    inward pointing 5
theorem
    0-Fredholm operators 50, 54
    0-Sobolev spaces 48
    L^2-boundedness
        0-calculus 46
        weighted 0-calculus 47
    Ψ^*-completion
        0-calculus 71
        b-c-calculus 71
    b-c-Fredholm operators 24
    characterization of reduced normal operator 38
    Jacobson topology 66
    properties of b-c-calculus 23
top face 5
triple space
    b-triple space 8
    edge triple space 7

extended 0-triple space........8

vector fields
    0-vector fields...........ix, xi, 3
    b-vector fields.............x, xv
    b-c-vector fields..............19
    cusp-vector fields..............x
    edge-vector fields..............x
    linear vector fields...........73

weight system.................16, 22
weighted 0-Sobolev spaces........47
weighted b-c-Sobolev spaces.......24

Editorial Information

To be published in the *Memoirs*, a paper must be correct, new, nontrivial, and significant. Further, it must be well written and of interest to a substantial number of mathematicians. Piecemeal results, such as an inconclusive step toward an unproved major theorem or a minor variation on a known result, are in general not acceptable for publication. Papers appearing in *Memoirs* are generally longer than those appearing in *Transactions*, which shares the same editorial committee.

As of February 1, 2003, the backlog for this journal was approximately 3 volumes. This estimate is the result of dividing the number of manuscripts for this journal in the Providence office that have not yet gone to the printer on the above date by the average number of monographs per volume over the previous twelve months, reduced by the number of volumes published in four months (the time necessary for preparing a volume for the printer). (There are 6 volumes per year, each containing at least 4 numbers.)

A Consent to Publish and Copyright Agreement is required before a paper will be published in the *Memoirs*. After a paper is accepted for publication, the Providence office will send a Consent to Publish and Copyright Agreement to all authors of the paper. By submitting a paper to the *Memoirs*, authors certify that the results have not been submitted to nor are they under consideration for publication by another journal, conference proceedings, or similar publication.

Information for Authors

Memoirs are printed from camera copy fully prepared by the author. This means that the finished book will look exactly like the copy submitted.

The paper must contain a *descriptive title* and an *abstract* that summarizes the article in language suitable for workers in the general field (algebra, analysis, etc.). The *descriptive title* should be short, but informative; useless or vague phrases such as "some remarks about" or "concerning" should be avoided. The *abstract* should be at least one complete sentence, and at most 300 words. Included with the footnotes to the paper should be the 2000 *Mathematics Subject Classification* representing the primary and secondary subjects of the article. The classifications are accessible from www.ams.org/msc/. The list of classifications is also available in print starting with the 1999 annual index of *Mathematical Reviews*. The Mathematics Subject Classification footnote may be followed by a list of *key words and phrases* describing the subject matter of the article and taken from it. Journal abbreviations used in bibliographies are listed in the latest *Mathematical Reviews* annual index. The series abbreviations are also accessible from www.ams.org/publications/. To help in preparing and verifying references, the AMS offers MR Lookup, a Reference Tool for Linking, at www.ams.org/mrlookup/. When the manuscript is submitted, authors should supply the editor with electronic addresses if available. These will be printed after the postal address at the end of the article.

Electronically prepared manuscripts. The AMS encourages electronically prepared manuscripts, with a strong preference for $\mathcal{AMS}$-LaTeX. To this end, the Society has prepared $\mathcal{AMS}$-LaTeX author packages for each AMS publication. Author packages include instructions for preparing electronic manuscripts, the *AMS Author Handbook*, samples, and a style file that generates the particular design specifications of that publication series. Though $\mathcal{AMS}$-LaTeX is the highly preferred format of TeX, author packages are also available in $\mathcal{AMS}$-TeX.

Authors may retrieve an author package from e-MATH starting from `www.ams.org/tex/` or via FTP to `ftp.ams.org` (login as `anonymous`, enter username as password, and type `cd pub/author-info`). The *AMS Author Handbook* and the *Instruction Manual* are available in PDF format following the author packages link from `www.ams.org/tex/`. The author package can be obtained free of charge by sending email to `pub@ams.org` (Internet) or from the Publication Division, American Mathematical Society, P.O. Box 6248, Providence, RI 02940-6248. When requesting an author package, please specify $\mathcal{AMS}$-LATEX or $\mathcal{AMS}$-TEX, Macintosh or IBM (3.5) format, and the publication in which your paper will appear. Please be sure to include your complete mailing address.

Sending electronic files. After acceptance, the source file(s) should be sent to the Providence office (this includes any TEX source file, any graphics files, and the DVI or PostScript file).

Before sending the source file, be sure you have proofread your paper carefully. The files you send must be the EXACT files used to generate the proof copy that was accepted for publication. For all publications, authors are required to send a printed copy of their paper, which exactly matches the copy approved for publication, along with any graphics that will appear in the paper.

TEX files may be submitted by email, FTP, or on diskette. The DVI file(s) and PostScript files should be submitted only by FTP or on diskette unless they are encoded properly to submit through email. (DVI files are binary and PostScript files tend to be very large.)

Electronically prepared manuscripts can be sent via email to `pub-submit@ams.org` (Internet). The subject line of the message should include the publication code to identify it as a Memoir. TEX source files, DVI files, and PostScript files can be transferred over the Internet by FTP to the Internet node `e-math.ams.org` (130.44.1.100).

Electronic graphics. Comprehensive instructions on preparing graphics are available at `www.ams.org/jourhtml/graphics.html`. A few of the major requirements are given here.

Submit files for graphics as EPS (Encapsulated PostScript) files. This includes graphics originated via a graphics application as well as scanned photographs or other computer-generated images. If this is not possible, TIFF files are acceptable as long as they can be opened in Adobe Photoshop or Illustrator. No matter what method was used to produce the graphic, it is necessary to provide a paper copy to the AMS.

Authors using graphics packages for the creation of electronic art should also avoid the use of any lines thinner than 0.5 points in width. Many graphics packages allow the user to specify a "hairline" for a very thin line. Hairlines often look acceptable when proofed on a typical laser printer. However, when produced on a high-resolution laser imagesetter, hairlines become nearly invisible and will be lost entirely in the final printing process.

Screens should be set to values between 15% and 85%. Screens which fall outside of this range are too light or too dark to print correctly. Variations of screens within a graphic should be no less than 10%.

Inquiries. Any inquiries concerning a paper that has been accepted for publication should be sent directly to the Electronic Prepress Department, American Mathematical Society, P. O. Box 6248, Providence, RI 02940-6248.

Editors

This journal is designed particularly for long research papers, normally at least 80 pages in length, and groups of cognate papers in pure and applied mathematics. Papers intended for publication in the *Memoirs* should be addressed to one of the following editors. In principle the Memoirs welcomes electronic submissions, and some of the editors, those whose names appear below with an asterisk (*), have indicated that they prefer them. However, editors reserve the right to request hard copies after papers have been submitted electronically. Authors are advised to make preliminary email inquiries to editors about whether they are likely to be able to handle submissions in a particular electronic form.

Algebra to KAREN E. SMITH, Department of Mathematics, University of Michigan, 525 University, Suite 2832, Ann Arbor, MI 48109-1109; email: `kesmith@lsa.umich.edu`

Algebraic geometry to DAN ABRAMOVICH, Department of Mathematics, Boston University, 111 Cummington Street, Boston, MA 02215; e-mail: `abrmovic@bu.edu`

Algebraic topology and cohomology of groups to STEWART PRIDDY, Department of Mathematics, Northwestern University, 2033 Sheridan Road, Evanston, IL 60208-2730; email: `priddy@math.nwu.edu`

Combinatorics and Lie theory to SERGEY FOMIN, Department of Mathematics, University of Michigan, Ann Arbor, Michigan 48109-1109; email: `fomin@umich.edu`

Complex analysis and complex geometry to DUONG H. PHONG, Department of Mathematics, Columbia University, 2990 Broadway, New York, NY 10027-0029; email: `phong@math.columbia.edu`

*__Differential geometry and global analysis__ to LISA C. JEFFREY, Department of Mathematics, University of Toronto, 100 St. George St., Toronto, ON Canada M5S 3G3; email: `jeffrey@math.toronto.edu`

Dynamical systems and ergodic theory to ROBERT F. WILLIAMS, Department of Mathematics, University of Texas, Austin, Texas 78712-1082; email: `bob@math.utexas.edu`

*__Geometric analysis__ to TOBIAS COLDING, Courant Institute, New York University, 251 Mercer Street, New York, NY 10012; email: `colding@cims.nyu.edu`

Geometric topology, knot theory and hyperbolic geometry to ABIGAIL A. THOMPSON, Department of Mathematics, University of California, Davis, Davis, CA 95616-5224; email: `thompson@math.ucdavis.edu`

Harmonic analysis, representation theory, and Lie theory to ROBERT J. STANTON, Department of Mathematics, The Ohio State University, 231 West 18th Avenue, Columbus, OH 43210-1174; email: `stanton@math.ohio-state.edu`

*__Logic__ to THEODORE SLAMAN, Department of Mathematics, University of California, Berkeley, CA 94720-3840; email: `slaman@math.berkeley.edu`

Number theory to HAROLD G. DIAMOND, Department of Mathematics, University of Illinois, 1409 W. Green St., Urbana, IL 61801-2917; email: `diamond@math.uiuc.edu`

*__Ordinary differential equations, and applied mathematics__ to PETER W. BATES, Department of Mathematics, Michigan State University, East Lansing, MI 48824-1027; email: `peter@math.msu.edu`

*__Partial differential equations__ to PATRICIA E. BAUMAN, Department of Mathematics, Purdue University, West Lafayette, IN 47907-1395' email: `bauman@math.purdue.edu`

*__Probability and statistics__ to KRZYSZTOF BURDZY, Department of Mathematics, University of Washington, Box 354350, Seattle, Washington 98195-4350; email: `burdzy@math.washington.edu`

Real analysis and partial differential equations to DANIEL TATARU, Department of Mathematics, University of California, Berkeley, Berkeley, CA 94720; email: `tataru@math.berkeley.edu`

All other communications to the editors should be addressed to the Managing Editor, WILLIAM BECKNER, Department of Mathematics, University of Texas, Austin, TX 78712-1082; email: `beckner@math.utexas.edu`.

Titles in This Series

777 **Robert Lauter,** Pseudodifferential analysis on conformally compact spaces, 2003

776 **U. Haagerup, H. P. Rosenthal, and F. A. Sukochev,** Banach embedding properties of non-commutative L^p-spaces, 2003

775 **P. Lochak, J.-P. Marco, and D. Sauzin,** On the splitting of invariant manifolds in multidimensional near-integrable Hamiltonian systems, 2003

774 **Kai A. Behrend,** Derived ℓ-adic categories for algebraic stacks, 2003

773 **Robert M. Guralnick, Peter Müller, and Jan Saxl,** The rational function analogue of a question of Schur and exceptionality of permutation representations, 2003

772 **Katrina Barron,** The moduli space of $N = 1$ superspheres with tubes and the sewing operation, 2003

771 **Shigenori Matsumoto,** Affine flows on 3-manifolds, 2003

770 **W. N. Everitt and L. Markus,** Elliptic partial differential operators and symplectic algebra, 2003

769 **Jie Wu,** Homotopy theory of the suspensions of the projective plane, 2003

768 **R. Höpfner and E. Löcherbach,** Limit theorems for null recurrent Markov processes, 2003

767 **Po Hu,** S-modules in the category of schemes, 2003

766 **Su Gao and Alexander S. Kechris,** On the classification of Polish metric spaces up to isometry, 2003

765 **Robert Bieri and Ross Geoghegan,** Connectivity properties of group actions on non-positively curved spaces, 2003

764 **J. Spandaw,** Noether-Lefschetz problems for degeneracy loci, 2003

763 **Yasuyuki Kachi and Eiichi Sato,** Segre's reflexivity and an inductive characterization os hyperquadrics, 2002

762 **Leiba Rodman, Ilya M. Spitkovsky, and Hugo Woerdeman,** Abstract band method via factorization, positive and band extensions of multivariable almost periodic matrix functions, and spectral estimation, 2002

761 **Oliver Druet and Emmanuel Hebey,** The AB program in geometric analysis : Sharp Sobolev inequalities and related problems, 2002

760 **Markus Banagl,** Extending intersection homology type invariants to non-Witt spaces, 2002

759 **Donald M. Davis,** From representation theory to homotopy groups, 2002

758 **Alan Forrest, John Hunton, and Johannes Kellendonk,** Topological invariants for projection method patterns, 2002

757 **Douglas Bowman,** q-difference operators, orthogonal polynomials, and symmetric expansions, 2002

756 **José Ignacio Cogolludo-Agustín,** Topological invariants of the complement to arrangements of rational plane curves, 2002

755 **M. A. Mandell and J. P. May,** Equivariant orthogonal spectra and S-modules, 2002

754 **Edward L. Green, Idun Reiten, and Øyvind Solberg,** Dualities on generalized Koszul algebras, 2002

753 **Daniel Panazzolo,** Desingularization of nilpotent singularities in families of planar vector fields, 2002

752 **Linus Kramer,** Homogeneous spaces, Tits buildings, and isoparametric hypersurfaces, 2002

For a complete list of titles in this series, visit the
AMS Bookstore at **www.ams.org/bookstore/**.